AF822120

Immunologic Research

A Selective Reference to Current Research and Practice

Formerly 'Survey of Immunologic Research'

KARGER

Printed in Switzerland
on acid-free paper by
Thür AG Offsetdruck, Pratteln

Appears quarterly:
1 volume per year (4 issues)

Immunologic Research

Submission: Manuscripts should be addressed to:

Julius M. Cruse, MD, PhD
Department of Pathology
Immunopathology Section
University of Mississippi Medical Center
2500 North State Street
Jackson, MS 39216 (USA)

Manuscripts should be submitted in triplicate (with three sets of illustrations of which one is an original), typewritten double-spaced on one side of the paper, with a wide margin.

Conditions: Both original and review papers are considered. Manuscripts must be written in English, and are all subject to editorial review. Manuscripts are received with the explicit understanding that they are not submitted for simultaneous consideration by any other publication. Submission of an article for publication implies the transfer of the copyright from the author to the publisher upon acceptance. Accepted papers become the permanent property of 'Immunologic Research' and may not be reproduced by any means, in whole or in part, without the written consent of the publisher. It is the author's responsibility to obtain permission to reproduce illustrations, tables, etc. from other publications.

Arrangement:
Title page: The first page of each paper should indicate the title (main title underlined), the authors' names, and the institute where the work was conducted. A short title for use as running head as well as the full address of the author to whom correspondence should be sent are also required.
Full address: The exact postal address complete with postal code must be given at the bottom of the title page. Please also supply Phone, Fax, and E-mail numbers.
Key words: For indexing purposes, a list of 3–10 key words in English is essential.
Abstract: Each paper needs an abstract of up to 10 lines.
Small type: Paragraphs which can or must be set in smaller type (case histories, test methods, etc.) should be indicated with a 'p' (petit) in the margin on the left-hand side.
Footnotes: Avoid footnotes. When essential, they are numbered consecutively and typed at the foot of the appropriate page.
Tables and illustrations: Tables and illustrations (both numbered in Arabic numerals) should be prepared on separate sheets. Tables require a heading and figures a legend, also prepared on a separate sheet. For the reproduction of illustrations, only good drawings and original photographs can be accepted; negatives or photocopies cannot be used. When possible, group several illustrations on one block for reproduction (max. size 144 × 196 mm) or provide crop marks. On the back of each illustration, indicate its number, the author's name, and 'top'.

Color illustrations: Up to 6 color illustrations per page can now be integrated with the text at the reduced price of SFr. 600.–/US$ 400.00 per page. Color illustrations are reproduced at the author's expense.

References: In the text identify references by Arabic numerals [in square brackets]. Material submitted for publication but not yet accepted should be noted as 'unpublished data' and not be included in the reference list. The list of references should include only those publications which are cited in the text. Do not alphabetize; number references in the order in which they are first mentioned in the text. The surnames of the authors followed by initials should be given. There should be no punctuation other than a comma to separate the authors. Cite all authors, 'et al' is not sufficient. Abbreviate journal names according to the Index Medicus system. (Also see International Committee of Medical Journal Editors: Uniform requirements for manuscripts submitted to biomedical journals. Br Med J 1988;296:401–405.)
Examples
(a) Papers published in periodicals: Kauffman HF, van der Heide S, Beaumont F, Blok H, de Vries K: Class-specific antibody determination against *Aspergillus fumigatus* by means of the enzyme-linked immunosorbent assay. III. Comparative study: IgG, IgA, IgM ELISA titers, precipitating antibodies and IgE binding after fractionation of the antigen. Int Arch Allergy Appl Immunol 1986;80:300–306.
(b) Monographs: Matthews DE, Farewell VT: Using and Understanding Medical Statistics. Basel, Karger, 1985.
(c) Edited books: Hardy WD Jr, Essex M: FeLV-induced feline acquired immune deficiency syndrome: A model for human AIDS; in Klein E (ed): Acquired Immunodeficiency Syndrome. Prog Allergy. Basel, Karger, 1986, vol 37, pp 353–376.
Further details: Authors may consult our leaflet 'Rules for the Preparation of Manuscripts', available on request from the Publisher, S. Karger AG, P.O. Box, CH-4009 Basel (Switzerland).

Page charges: Page charges may be levied if the allotted size of the manuscript, as stipulated by the editors of the journal, is exceeded. Each additional complete or partial page is charged to the author at SFr. 245.– / US$ 196.00. 2 manuscript pages (including tables, illustrations and references) are equal to approximately one printed page.

Galley proofs: Unless indicated otherwise, galley proofs are sent to the first-named author and should be returned with the least possible delay. Alterations made in galley proofs, other than the correction of printer's errors, are charged to the author. No page proofs are supplied.

Reprints: Order forms and a price list are sent with the galley proofs. Orders submitted after the issue is printed are subject to considerably higher prices. Allow five weeks from date of publication for delivery of reprints.

KARGER

General Information

Immunologic Research

Publication data: 'Immunologic Research' is published 4 times annually. Volume 13 with 4 issues appears in 1994.

Subscription rates: Subscriptions run for a full calendar year. Prices are given per volume.

Personal subscription:

Subscription	*+ postage and handling*	
SFr. 150.–	+ SFr. 6.40	Europe
SFr. 150.–	+ SFr. 9.60	Overseas
US$ 100.00	+ US$ 6.40	

(Must be in the name of, billed to, and paid by an individual. Order must be marked 'personal subscription'.)

Institutional subscription:

Subscription	*+ postage and handling*	
SFr. 375.–	+ SFr. 16.–	Europe
SFr. 375.–	+ SFr. 24.–	Overseas
US$ 250.00	+ US$ 16.00	

Airmail postage: SFr. 24.–, US$ 20.00. (Extra per volume.)

Discount subscription prices: American Association of Immunologist.

Microform edition: SFr. 187.50, US$ 150.00. (Available exclusively to subscribers of the paper edition.)

Subscription orders: Orders can be placed at agencies, bookstores, directly with the Publisher S. Karger AG, P.O. Box, CH – 4009 Basel (Switzerland) or with any of the following distributors:

Deutschland: S. Karger GmbH, Postfach
D – 79095 Freiburg

France: Librairie Luginbühl
36, bd de Latour-Maubourg
F–75007 Paris

Great Britain, Ireland: S. Karger Publishers
11 Ravenscroft Road
Beckenham, Kent BR3 4TP

USA: S. Karger Publishers, Inc.
26 West Avon Road, P.O. Box 529
Farmington, CT 06085

India, Bangladesh, Sri Lanka:
S. Karger India Subscription Agency
B – 5/132 Safdarjung Enclave
New Delhi-110029

Japan: Katakura Libri, Inc.
36-9 Hongo 3-chome, Bunkyo-ku
Tokyo 113

Change of address: Both old and new address should be sent to the subscription source.

Single issues and back volumes: Information can be obtained through the Publisher.

Advertising: Correspondence and rate requests should be addressed to the Publisher.

Bibliographic indices: This journal is regularly listed in bibliographic services, including *Current Contents®*.

KARGER

Summarizes recent advances in this fast-moving field

Selective Immunosuppression: Basic Concepts and Clinical Applications

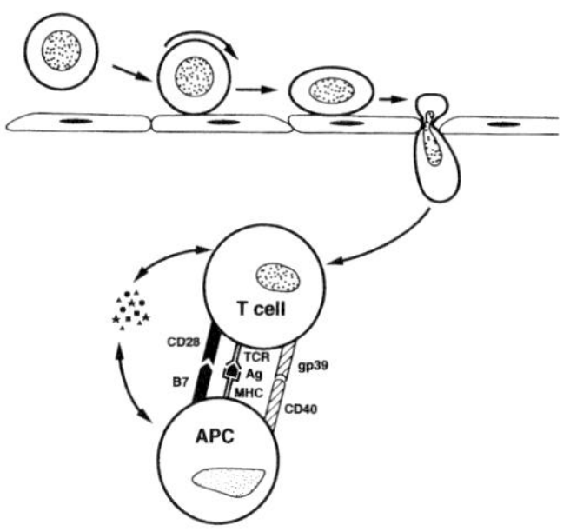

Editor: Luciano Adorini

Induction of selective immunosuppression is a challenging research topic, and enough is now known about the immune system to permit new ideas on immunoregulation to be applied clinically. This book provides an authoritative update of developments in this fast-moving field from both a basic and clinical perspective.

Leading immunologists highlight appropriate targets for selective immunotherapy of autoimmune diseases, allograft rejection and allergies. Strategies based on autoantigen, MHC, TCR, cytokines, coreceptors, adhesion molecules and Th1/Th2 balance are analyzed in detail in order to understand the underlying mechanisms and to explore their possible impact on the formulation of increasingly more selective immunotherapies.

Immunologists and immunopharmacologists will find this book of particular interest, and it should also appeal to molecular and cell biologists.

Contents

Chemical Immunology, Vol. 60
Series Editors: Adorini, L. (Milan); Arai, K. (Tokyo); Capra, J.D. (Dallas, Tex.); Ishizaka, K. (La Jolla, Calif.); Waldmann, H. (Cambridge); Waksman, B.H. (New York, N.Y.)
ISSN 1015–0145
Fields of Interest: Immunology; Allergy, Biochemistry, Diabetes, Immunosuppression, Oncology, Pharmacology, Rheumatology, Transplantation

Selective Immunosuppression: Basic Concepts and Clinical Applications
Editor: Adorini, L. (Milan)
X + 166 p., 22 fig., 7 tab., hard cover, 1995
Institutions: SFr. 174.– / DM 208.– / US$ 151.50
Individuals: SFr. 87.– / DM 104.– / US$ 75.75
Prices subject to change
DM price for Germany, US$ price for USA only
ISBN 3–8055–6034–6

KI 95508

KARGER

Immunology: A University of Chicago Perspective

Guest Editor
J.A. Bluestone, Chicago, Ill., USA

21 figures, 1 table, 1995

KARGER

S. Karger
Medical and Scientific Publishers
Basel · Freiburg · Paris · London
New York · New Delhi · Bangkok
Singapore · Tokyo · Sydney

Drug Dosage
The authors and the publisher have exerted every effort to ensure that drug selection and dosage set forth in this text are in accord with current recommendations and practice at the time of publication. However, in view of ongoing research, changes in government regulations, and the constant flow of information relating to drug therapy and drug reactions, the reader is urged to check the package insert for each drug for any change in indications and dosage and for added warnings and precautions. This is particularly important when the recommended agent is a new and/or infrequently employed drug.

Printed in Switzerland on acid-free paper by
Thür AG Offsetdruck, Pratteln
ISBN 3-8055-6142-3

Contents

Immunol Res 1994;13:213–214

Jeffrey A. Bluestone, PhD

Chairman,
Committee on Immunology

Introduction

As biology and medicine enter a new era of discovery, immunology has become a primary locus for research into diseases as disparate as diabetes and AIDS. The Committee on Immunology at the University of Chicago is at the forefront of this biologic revolution. The Committee on Immunology was founded two decades ago as an outgrowth of the community of university immunologists' need for an organizational structure to bring together faculty members to build on a tradition of research excellence, and to encourage collaborations, new recruits, and discovery.

Immunology research at the University of Chicago dates to the first decade of the twentieth century when Dr. Alexis Carel, a Research Associate in the Department of Physiology, performed the first human tissue graft. His Nobel Prize-winning research began the field of transplant biology, a prominent area of research and clinical activity in our institution today. Dr. Howard Taylor Ricketts, the discoverer of the organisms that cause Rocky Mountain Spotted Fever and typhus, wrote one of the first immunology textbooks as an Assistant Professor of Pathology at the University of Chicago.

Dr. David Talmage ranks among the most prominent early immunologist on the faculty. On arriving at the University in 1952, he began studying antibody diversity with Drs. William Tooly and Lucy Taliaferro. In his historical account published in the *Annual Reviews of Immunology* (1986), he retells the story of how he arrived at the clonal selection theory of antibody formation published in the *Annual Reviews of Medicine* in 1957, the same year Burnet published his famous clonal selection paper. Needless to say, their work has had a profound influence on subsequent thinking about immune selection.

Following World War II, the University of Chicago entered a golden age of basic research. Dr. Paul Cannon, the third chairman of the Department of Pathology and a member of the National Academy of Sciences, researched the effects of protein depletion on immunity, a subject of great importance because of the widespread war-related malnutrition in Europe and Asia. Dr. Robert Wissler, who had done his thesis with Cannon, collaborated with Dr. Leon Jacobsen in early studies of the effects of radioactivity on hematopoiesis and the immune system. The Manhattan Project had made the University of Chicago (and its Argonne Cancer Research Hospital) a focus for this type of study, and Jacobsen encouraged Dr. Eugene Goldwasser to study hematopoiesis; Goldwasser's pioneering work led to his discovery of the hormone erythropoietin in 1979. Similarly, Dr. Frank Fitch became interested in the study of hematopoiesis following lethal irradiation. In collaboration with Jacobsen and Wissler, Fitch showed that shielding the spleen during lethal irradiation resulted in a full reconstitution and survival, suggesting that repopulation of hematopoietic lineages could be accomplished by cells residing in the spleen. This work resulted in Fitch's first publication which led to a career that, without doubt, has placed him among the world's preeminent immunologists.

Immunology research was not limited to the main campus at the University of Chicago. LaRabida Hospital is a chronic care children's hospital affiliated with the university that, in the 1940s and 1950s was devoted almost exclusively to the care of children with rheumatic fever. Drs. Eugene Fox and Ed Cohen were based at LaRabida working on streptococcal immunity. In the 1970s, Heinz Kohler, Hans Schreiber, Mike Loken and Jose Quintans joined a very strong immunology program at LaRabida led by Dr. Don Rowley. I joined the immunology group at the University of Chicago in 1987, as part of a new cohort of immunologists that included Drs. Ursula Storb and Jim Miller. In addition, Don Rowley, Hans Schreiber, and Jose Quintans moved to the main campus from LaRabida Hospital. The consolidation of immunology on the main campus moved the immunology program into a new era of growth. During the past 8 years, more than 10 basic immunologists have joined our faculty, among them Drs. Jeffrey Leiden, Andrea Sant, Marcus Clark and Craig Thompson through continued support of the University of Chicago and the Gwen Knapp Center for Lupus and Immunology Research. Ongoing research areas of committee interest include: T and B cell development, T and B cell signalling, antigen presentation, transplant immunology, and autoimmune diseases.

The program is vibrant and exciting. The long history of immunology at the University of Chicago has led to a stable, highly interactive group of basic and clinical researchers. We hope that these review articles offer some real flavor for the research done in our laboratories. More important, I hope that our efforts will contribute to a greater understanding of the dynamics of the immune system, as well as its development, regulation, and role in disease.

Immunol Res 1994;13:215–233

David C. Cronin, II[a]
David W. Lancki[a]
Frank W. Fitch[a,b]

[a] Committee on Immunology, Department of Pathology, and
[b] Ben May Institute, University of Chicago, Ill., USA

Requirements for Activation of CD8+ Murine T Cells

I. Development of Cytolytic Activity

Key Words
Cytolysis
CD8+ T cells, activation
T cell proliferation
α/β TCR transgene

Abstract

Cytolytic effector function fails to develop if proliferation of allospecific cytolytic T lymphocyte precursors is inhibited, but the requirements for generation of cytolytic activity have not been fully defined. In contrast, the cytolytic effector function of cytolytic T lymphocyte clones does not change during the cell cycle, and the level of cytolytic activity is independent of cellular proliferation. The requirement for proliferation by primary responding populations may reflect the need for clonal expansion of a few inherently cytolytic effector cells in order to reach a threshold number which can readily be detected in conventional cytolytic assays. Alternatively, proliferation may be required for cytolytic T lymphocyte precursors to differentiate into mature, functional cytolytic cells. Using CD8+ T cells which express an antigen-specific transgenic α/β T cell receptor, we have studied the requirements for acquisition of cytolytic capacity. Stimulation of the T cell receptor alone appears to be sufficient to render naive, CD8+ transgenic T cells sensitive to the growth effects of interleukin-2 (IL-2), and in some circumstances to interleukin-4 (IL-4), but not to induce either lymphokine production or cytolytic activity. Costimulatory molecules expressed by allogenic stimulating cells appear to be required for lymphokine production, and CD8+ transgenic T cells initially appear to secrete only IL-2 and interferon-γ. Stimulation of the T cell receptor of naive, CD8+ transgenic T cells appears to induce cytolytic activity only if cell proliferation occurs, either in response to IL-2 produced by the stimulated cells themselves when costimulatory molecules are present, or to IL-2 or IL-4 from exogenous sources if costimulatory molecules are absent.

Introduction

The requirements for generation of allospecific cytolytic T lymphocytes (CTL) from CTL precursors have not been fully defined. A number of earlier reports indicated that cytolytic effector function fails to develop if proliferation of the responding cells is inhibited. However, these studies have employed mixed lymphocyte culture prepared with heterogeneous populations of murine lymphoid cells as the model system [1–5]. A dependence upon proliferation for the generation of cytolytic activity in vivo also has been demon-

Dr. Frank W. Fitch
Department of Pathology, MC1089
University of Chicago
5841 S. Maryland Avenue
Chicago, IL 60637 (USA)

0257–277X/94/
0134–0215$8.00/0

strated by Denizot et al. [6]. Lymphokines also appear to be involved in the development of cytolytic activity, but their role has not been fully characterized. An important function of lymphokines appears to relate to the promotion of CTL replication.

Optimal generation of CTL specific for allogeneic class I MHC determinants usually requires the additional T cell recognition of class II MHC determinants by CD4+ helper T cells [7]. Interleukin (IL)-2 produced by CD4+ T cells generally is required for the proliferation of CD8+ CTL [8]. Other lymphokines, including IL-4 [9], IL-6 [10], and IL-10 [11], may influence the development of cytolytic activity. Antigen-independent proliferation was not sufficient to induce cytolytic activity in CD8+ T cells in some systems [12, 13]. When concanavalin A-stimulated, Lyt-2+ murine splenic T cells were also exposed to various lymphokine-containing supernatants, the extent of proliferation did not correlate with the magnitude of the cytolytic response generated, and it was concluded that although IL-2 could cause replication of CTL precursors, other factors were required for differentiation into CTL [13].

Pharmacologic agents that interfere with cell division have been used in many of the studies documenting the requirement of cell replication for generation of cytolytic activity. However, it is difficult in heterogeneous systems to determine the mechanisms by which mitotic inhibitors interfere with the development of cytolytic activity. Such inhibitors may act either on the CD4+ helper cells to limit their ability to provide help for the CD8+ CTL precursors, or they may act directly on the CD8+ T cells. In addition, the low frequency of cells which can respond to any given antigen makes it difficult to determine if the failure to detect antigen-specific cytolytic activity in unstimulated T cell populations is due to an insufficient number of T cells that have inherent cytolytic function or to a requirement for proliferation during the differentiation of noncytolytic CTL precursors into cytolytic cells. Activation of the majority of T cells with lectins has been used in some studies in an attempt to overcome the limitations of low precursor frequency [3], but such mitogens may have effects in addition to those associated with T cell receptor (TCR) stimulation.

In an effort to define more precisely the relationships between TCR stimulation, proliferation, and the acquisition of cytolytic capability and other T lymphocyte functions, we have utilized CD8+ T cells which express an antigen-specific, transgenic α/β-TCR [14]. The development of the cytolytic activity and the ability to secrete lymphokines in this system involves responding cells which express clonally a single TCR and are present at a high frequency. The results that we have obtained indicate that TCR stimulation followed by cell proliferation is necessary and sufficient for the development of cytolytic activity. Cells in which TCR stimulation is not followed by cell replication do not become cytolytic.

Materials and Methods

Animals

Animals were housed in microisolator caging and maintained in laminar flow units in a barrier-type facility. Female DBA/2 mice were purchased from the Jackson Laboratories, Bar Harbor, Me., USA. Male and female α/β-TCR transgenic C57BL/6 mice developed by Dr. Dennis Loh [14] were bred in the animal facility using stock obtained from Dr. Loh at Washington University, St. Louis, Mo., USA.

Culture Media

T cells were grown in culture medium consisting of DMEM plus 10% heat-inactivated fetal calf serum, 10 mM 3(N-morpholino)propane sulfonic acid, 5×10^{-5} M 2-mercaptoethanol, 100 U/ml penicillin, 100 μg/ml streptomycin, and additional amino acids [15]. The various indicator cell lines used for lymphokine assays, as well as other cells as indicated, were grown in the same complete medium containing 10% fetal calf serum.

Reagents

Phorbol myristate acetate (PMA), aphidicolin, hydroxyurea (H 8627), nocodazole (methyl[5-(2-thienylcarbonyl)-1H-benzimidazol-2-yl]carbamate, M 1404), and mitomycin C (M 0503) were all purchased from Sigma Chemical Co. (St. Louis, Mo., USA). Mimosine (M8, 761-4) was purchased from Aldrich Chemical Co. (Milwaukee, Wisc., USA). Colcemide and Dulbecco's phosphate-buffered saline (DPBS), without calcium and magnesium (catalog No. 7732-18-5) were purchased from Gibco (Grand Island, N.Y., USA).

Recombinant Lymphokines

Recombinant murine IL-2 and IL-4 were purchased from Genzyme Corp. (Boston, Mass., USA). Recombinant human IL-2 was generously provided by Cetus Corp. (Emeryville, Calif., USA). Human rIL-2 was used for derivation and maintenance of T cell clones; murine rIL-2 was used as a standard in the IL-2 bioassay. All concentrations are expressed in terms of units per milliliter as defined by the supplier of each of the lymphokines.

Monoclonal Antibodies (mAb)

Hybridoma lines 145-2C11 (anti-CD3-ε) [16], 11B11 (anti-IL-4) [17], S4B6 (anti-IL-2) [18], and 1B2 (anti-clonotypic mAb reactive with clone 2C TCR) [19] were kindly provided by Dr. J. Bluestone (University of Chicago, Ill., USA), Dr. W. Paul (NIH, Bethesda, Md., USA), Dr. T. Mosmann (DNAX, Palo Alto, Calif., USA), and Dr. Dennis Loh (Washington University, St. Louis, Mo., USA), respectively. Hybridoma line R46A2 (anti-IFN-γ) [20] was obtained from the American Type Culture Collection (Rockville, Md., USA). Anti-CD3ε-FITC, anti-CD4-FITC, and anti-CD8-FITC mAb were purchased from PharMingen (San Diego, Calif., USA).

Supernatants were obtained from cultures of the following hybridomas and used as sources of mAb: FD441.8 (anti-LFA-1) [21]; 2.43 (anti-CD8) [22]; GK1.5 (anti-CD4) [23]; AT83A (anti-Thy-1.2) [24]; 2.4G2 (anti-FcγRII) [25], and PC61.5 (anti-IL-2 receptor) [26].

T Cell Clones and Cell Lines

The L^d-specific CTL clone L3 is CD8+, CD4–, Thy-1.2+ and secretes interferon-γ (IFN-γ) but not IL-2 or IL-4 [15, 27]. Cells were maintained by weekly stimulation with DBA/2 spleen cells in the presence of 10 U/ml recombinant human IL-2. The $H\text{-}2^d$, murine mastocytoma cell line, P815 (ATCC TIB 64), and EL-4 cells ($H\text{-}2^b$) were grown and used as target cells in cytolytic assays, and as a source of alloantigen for the transgenic T cells in assays of proliferation and cytokine secretion. When used to stimulate CD4-depleted, transgenic splenocytes, P815 mastocytoma cells were incubated with mitomycin C (Sigma) at a final concentration of 100 μg/ml for 90 min at 37 °C. The cells were then washed with 5 vol of complete media, incubated in complete medium for an additional 60 min, washed with 3 vol of complete media, and resuspended in complete medium. Mastocytoma cells were used at a concentration of 9×10^5 cells/well in 24-well cultures and at a concentration of 3×10^5 cells/well in 96-well cultures.

Preparations of the Naive CD8+ T Cell Populations

Transgenic mice were injected intraperitoneally with 0.2 ml of a sterile ascites preparation of GK1.5 mAb (anti-CD4) having an immunoglobulin concentration of 3 mg/ml in order to deplete CD4+ T cells in vivo. After 4–5 days, spleens and/or lymph nodes were harvested and used to prepare naive CD4+-depleted lymphoid cell suspensions. These populations routinely contained less than 1% detectable CD4+ cells as determined by FACS analysis.

Purification of CD8+ T Cells

CD8+ T cells were purified by positive selection with the aid of a MACS column (Miltenyi Biotec Inc., Auburn, Calif., USA). Briefly, CD4+ cells were depleted in vivo as described above, and a single-cell suspension was prepared. Erythrocytes were removed using ammonium chloride lysis buffer, and dead cells were removed by Ficoll-Hypaque gradient centrifugation [28]. Cells were filtered through Nytex mesh to remove debris and resuspended in MACs buffer (5 m*M* EDTA, 0.01% sodium azide in phosphate-buffered saline, pH 7.2–7.4). Anti-Ly-2 microbeads (MACs 494-01) were added, and the mixture was incubated on ice for 20 min. Alternatively, cells were incubated with biotinylated anti-CD8 mAb (PharMingen), washed, and incubated on ice for 20 min after the addition of avidin-fluorescein isothiocyanate (FITC, Becton-Dickison). This cell suspension was washed with MACs buffer, and biotin-coated MACs magnetic beads were added at a volume of 5 μl of bead suspension per 10^6 CD8+ cells. The cells were incubated on ice for 5 min, then passed over the magnetized MACS column. The column was sequentially washed with 5 column volumes of MACs buffer. The column was then removed from the magnet and flushed with 30 ml of buffer solution. The flushed solution was collected and washed. FACS analysis showed that the final preparation contained more than 96% CD8+ transgenic T cells, and the remaining cells did not express CD3.

Preparation of T-Depleted Splenocytes

Single-cell suspensions were prepared from spleens, and erythrocytes were removed using ammonium chloride lysis buffer. Cells were then suspended in cold AT83A (anti-Thy-1) mAb culture supernatant [24] and incubated on ice for 30 min with frequent agitation (1×10^8 splenocytes/5 ml of AT83A supernatant). After resuspension in cold Cyto-Tox medium (Accurate Chemical Westbury, N.Y., USA; 2×10^8 splenocytes/9 ml), freshly reconstituted Low-Tox guinea pig complement (Accurate Chemical) was added, and the mixture was incubated at 37 °C for 45 min. Viable cells were isolated by Ficoll-Hypaque gradient centrifugation. This population was devoid of Thy1+, CD8+, CD4+ and CD3+ cells by FACS analysis.

Stimulation of T Cells with Immobilized Anti-CD3 mAb

Wells of 96-well flat-bottom culture plates (Costar 3596) were incubated overnight at 37 °C with 50 µl of goat antihamster antiserum (Organon Teknika Corp., West Chester, Pa., USA) at a concentration of 40 µg/ml in DPBS [29]. Wells were then washed twice with DPBS, and 50 µl of a 1:10 dilution of 145-2C11 hybridoma supernatant in DPBS was added. After incubation at 37 °C for 2 h or overnight at 4 °C, the plates were washed twice, and bulk or cloned T cells were added to the wells in a final volume of 0.2 ml. Supernatants were collected following incubation at 37 °C for 21–24 h, centrifuged to remove cells, and stored at 4 °C until assayed. For generation of large quantities of T cell supernatants, tissue culture dishes (Falcon 3002) were coated overnight at 37 °C with 1.5 ml of goat antihamster antiserum at a concentration of 40 µg/ml in DPBS, washed twice with 2 ml DPBS, and 1.5 ml of a 1:20 dilution of 145-2C11 hybridoma supernatant in DPBS was added. After incubation for 2 h, plates were washed twice, and T cells were added in a final volume of 4 ml. Supernatants were collected after incubation for 21–24 h, filtered through 0.2-µm filters (Nalgene 190-2020), and stored at 4 °C until assayed. For investigation of dose response to anti-CD3 antibody, 50 µl of 40 µg/ml goat antihamster antiserum was incubated overnight at 37 °C in each well of a 96-well flat-bottom plate. The wells were washed with DPBS, treated for 2 h with various dilutions of 145-2C11 supernatant, washed again, and T cells and the indicated reagents were added.

T Cell Proliferation Assays

CTL clone L3 cells were harvested from maintenance cultures after culture for 7–10 days, centrifuged over a Ficoll-Hypaque gradient, washed, and resuspended at a concentration of 1×10^6 cells/ml. Naive, CD4-depleted, transgenic splenocytes and purified CD8+ transgenic T cells were prepared as described and resuspended at a concentration of 1×10^6/ml. Each culture well of a 96-well microtiter plate (Costar 3596) or U-bottom microtiter plates (Linbro 76-013-05) received 2×10^4 purified T cells/well or 1×10^5 or 3×10^5 CD4-depleted splenocytes/well in 0.2 ml final volume along with the indicated concentrations of various reagents and stimulator populations. After various intervals, cultures were pulsed with 1 µCi/well of ^{3}H-TdR (New England Nuclear, Boston, Mass., USA) and harvested 13 h later onto glass filters using an PhD cell harvester (Cambridge Technology, Cambridge, Mass., USA). Later proliferation assays were harvested onto Packard 96-well UniFilter-96, GF/C plates (part No. 6005174) and counted on the Packard Top Count Microplate Scintillation Counter (Packard, Meriden, Conn., USA). Radioactivity was assessed by liquid scintillation spectrometry and expressed as counts per minute. If lymphokine and proliferation assays were performed using the same cultures, 100 µl of supernatant was removed prior to the thymidine pulse and used in the lymphokine assay.

In experiments using inhibitors of the cell cycle, naive, CD4-depleted transgenic splenocytes (2×10^6 cells) or L3 CTL (1×10^6) were cultured with 1×10^7 irradiated, T-depleted DBA/2 splenocytes and 5–10 U/ml of recombinant human IL-2. Cultures were prepared in a total volume of 1.5 ml of complete medium in 24-well tissue culture plates (Linbro). At the initiation of culture the following G1/S phase cell cycle blockers were added: aphidicolin 5 µg/ml, mimosine 200 µ*M*, and hydroxyurea 1 m*M*. The following M phase cell cycle inhibitors were used: colcemide 0.03 µg/ml and nocodazole 0.04 µg/ml. These concentrations were determined to be the lowest inhibitory concentrations, based on published reports and pilot experiments. After culture for 60 h, cells were gently resuspended and 200-µl aliquots of the cultures were transferred in triplicate to U-bottom microtiter plates and pulsed with 1 µCi/well of ^{3}H-thymidine for an additional 12 h. The wells were harvested as described, and the incorporated radioactivity expressed as the average counts per minute. Inhibition of cell proliferation was calculated as percent inhibition derived according to the formula:

$$\% \text{ inhibition} = 100 \times \frac{(\text{cpm inhibitor culture})}{(\text{cpm control culture})}$$

Cytologic Assay

Cell-mediated lysis was measured by a modification of the ^{51}Cr-release assay as described by Cerotini et al. [30]. Duplicate 3-fold serial dilutions of T cells were prepared in V-bottom microtiter plates (Linbro Catalog No. 76-223-05). Effector cells were incubated with 2,500 ^{51}Cr-labeled tumor target cells in a final volume of 0.2 ml for 5 h. Culture supernatant containing mAb 145-2C11 was included in redirected cytolytic assays. Plates were centrifuged at 250 *g*, and 0.1 ml of supernatant was collected and counted in a gamma scintillation counter. Lysis was calculated as follows:

$$\% \text{ specific lysis} = 100 \times \frac{(\text{experimental release} - \text{spontaneous release})}{(\text{maximum release} - \text{spontaneous release})}$$

Spontaneous release was determined by incubating labeled target cells in medium alone. Maximum release was determined following freeze-thaw or HCl-induced lysis of target cells. T cells were considered to exhibit lytic activity when ^{51}Cr-release values exceeded the spontaneous release values by 10%.

BLT Serine Esterase Assay

Total cellular content of N-α-benzyloxycarbonyl-*L*-lysin thiobenzyl (BTL) esterase in T cells (0.5–1 × 10^5 cells) was determined using a modification of the method of Coleman and Green [31]. Cells were centrifuged over a Ficoll-Hypaque density gradient, washed 3 times in DMEM, and then lysed in 200 μl of DMEM with 1% Triton X-100. Twenty microliters of this lysate was added to microtiter plates (Costar No. 3592). Assay solution (200 μl per well) consisting of a Tris-buffer (pH 7.2), with 0.22 μ*M* 5,5′-dithio-bis-(2-nitro-) benzoic acid and 0.20 μ*M* BLT (Calbiochem, Behring Diagnostic, La Jolla, Calif., USA) was then added. Plates were incubated (15–30 min) until a lysate of CTL clone L3 cells (used as a control standard) reached an absorbance level of 1.00 when read using a micro ELISA reader at test wavelength = 405 nm.

Lymphokine Assays

IFN-γ concentrations were determined using an ELISA [32] with modifications [33]; the reagents used in this assay were kindly provided by Dr. R. Schreiber (Washington University, St. Louis, Mo., USA). Activity in culture supernatants was compared to that in a standard (kindly provided by the NIAID, catalog No. Gg02-901-533). IL-2 and IL-4 concentrations were measured using a colorimetric 3-(4,5-dimethylthiazol-2-yl)-2,5-diphenyltetrazolium bromide (MTT, Sigma) assay [34]. Supernatants were tested for their ability to induce proliferation of the indicator cell line CTLL-2, which responds to both lymphokines. Monoclonal antibodies specific for IL-4 (11B11) [17] or IL-2 receptors (3C7 + PC61.5) [26, 35] and IL-2 (S4B6) [18] were used to confirm the identity of the T cell growth factors present in supernatants. In the presence or absence of mAb, 7.5 × 10^3 cells were incubated with 3-fold serial dilutions of culture supernatants at 37 °C for 24 h, MTT (25 μl at 2 mg/ml) was added to each microwell. Cultures were incubated for an additional 24 h at 37 °C; acidified 10% SDS was added to the microcultures, and the absorbance was read using a microtiter plate reader at a wavelength of 570 nm. Concentrations were determined by comparing samples to recombinant lymphokine standards.

Analysis of Cells by Flow Cytometry

Single-cell suspensions of splenocytes, various cell lines, and clones were prepared by established methods. Cells were suspended in FACS buffer (0.1% bovine serum albumin, 0.01% sodium azide in phosphate-buffered saline). Preincubation with mAb 2.4G2 (anti-Fc receptor) was carried out, prior to staining where indicated, to prevent binding of antibody to Fc receptors. All staining was performed in V-bottom microtiter plates (Linbro 76-321-05) at 4 °C. Cells (2–10 × 10^5) were incubated with the first step mAb or primary-conjugated antibodies for 15 min in a volume of 25 μl. Cells labeled with primary conjugated antibodies were washed with FACS buffer, resuspended in 300 μl and analyzed.

Those cells stained with the first step antibody (control mAb UC8-3B9 or anti-CD3 mAb) were washed in FACS buffer over a cushion of 25 μl fetal calf serum and centrifuged at 2,000 rpm for 5 min at 4 °C. Supernatants were aspirated and the cells were resuspended in 25 μl of second step antibody (1:20 dilution of affinity-purified, FITC-conjugated goat antihamster Ig; Jackson Immunological Research, West Grove, Pa., USA). Alternatively, cells were incubated directly with 1B2-FITC clonotypic TCR mAb. After incubation for 30 min at 4 °C, the cells were washed with FACS buffer and resuspended in 300 μl of buffer, separated into single-cell suspensions by filtering through a Nytex mesh, and analyzed. Propidium iodide was added to all samples, and analysis excluded dead cells.

Analysis was performed using a FACScan flow cytometer (Becton-Dickinson Immunocytometry Systems, Mountain View, Calif., USA) interfaced to a Hewlett-Packard 310 computer. Data analyses were performed using Consort-30 and Lysis II software. Logarithimically amplified fluorescence data were collected on 5,000–10,000 viable cells.

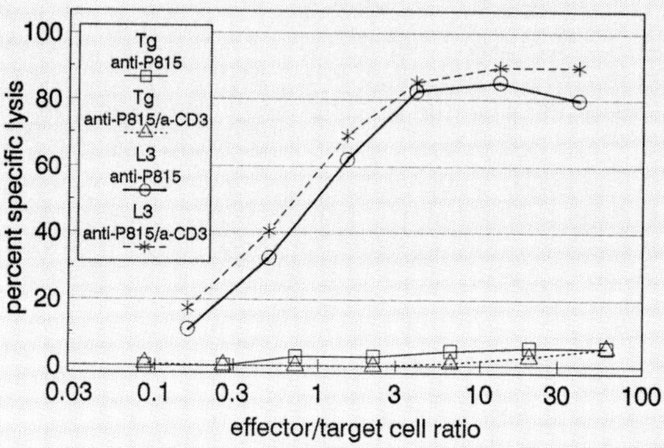

Fig. 1. Naive, CD4-depleted, transgenic splenocytes do not lyse target cells bearing the relevant L^d alloantigen. The lytic capacity of freshly prepared naive, CD4-depleted transgenic splenocytes was compared to that of the CTL clone L3. Freshly prepared CD4-depleted transgenic splenocytes and L3 cells, obtained from 7-day maintenance cultures, were incubated with ^{51}Cr-labeled P815 target cells (2.5×10^3 cells) with or without anti-CD3 mAb (1:8 dilution of hybridoma culture supernatant) for 5 h as described in 'Materials and Methods'. Spontaneous ^{51}Cr release in this experiment was 6.5%. These results are representative of those obtained in three similar experiments.

Results

Naive, CD4-Depleted, Transgenic Splenocytes Do Not Possess Inherent Cytolytic Capacity or Express BLT Esterase Activity

CD4-depleted spleen cells were prepared from α/β-TCR transgenic mice as described in 'Materials and Methods'. This splenocyte population contained 16–20% transgene-positive, CD8+ T cells and was devoid of CD4+ T cells as demonstrated by FACS analysis (data not shown). The cytolytic capacity of this naive CD4-depleted transgenic splenocyte population was investigated using alloantigen-bearing, P815 (H-2^d) mastocytoma target cells (fig. 1). Freshly isolated, CD4-depleted, transgenic splenocytes did not display cytolytic activity when tested in a 5-hour ^{51}Cr-release cytolytic assay, even at a high effector-to-target ratio (60:1). Furthermore, this CD4-depleted transgenic splenocyte population was unable to lyse targets in a redirected cytolytic assay which employed anti-CD3ε mAb (fig. 1). In contrast, CTL clone L3 cells obtained from maintenance cultures caused 50% specific lysis at an effector:target cell ratio of less than 1:1 (fig. 1).

To determine if freshly isolated CD4-depleted transgenic splenocytes contained serine esterases that have been implicated in CTL-mediated cytolysis, whole cell lysates were prepared from such cells and from CTL clone L3 cells obtained from maintenance culture and were assayed for the presence of BLT esterase activity as described in 'Materials

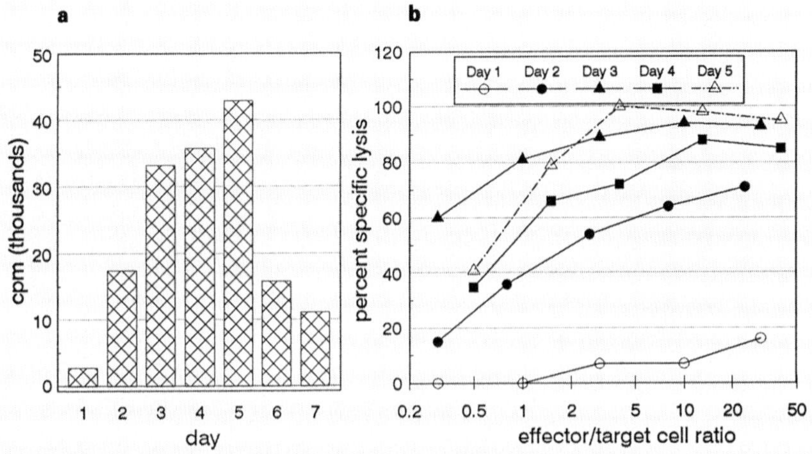

Fig. 2. a Time course of proliferation of naive, CD4-depleted, transgenic splenocytes stimulated with allogeneic, T-depleted DBA/2 splenocytes. Freshly prepared, CD4-depleted, transgenic splenocytes (1×10^5) were cultured in triplicate with irradiated, T-depleted DBA/2 splenocytes (1×10^6) in U-bottom microtiter plates as described in 'Materials and Methods'. Wells were pulsed with ^{3}H-thymidine for the final 12 h of the indicated culture period, harvested, and the incorported radioactivity expressed as the average counts per minute. These results are representative of those obtained at different time points in multiple experiments. ^{3}H-thymidine incorporation for unstimulated CD4-depleted, transgenic splenocytes ranged from 114 to 373 cpm; that for irradiated, T-depleted DBA/2 splenocytes ranged from 220 to 2,158 cpm. **b** Time course of development of cytolytic activity by naive, CD4-depleted, transgenic splenocytes stimulated with allogeneic, T-depleted DBA/2 splenocytes. Cultures prepared with CD4-depleted, transgenic cells (1×10^6) and irradiated, T-depleted DBA/2 splenocytes (6×10^6) as described in 'Materials and Methods' were collected after 24, 48, 72, 96, and 120 h in culture and combined with ^{51}Cr-labeled P815 targets in a 5-hour cytolytic assay as described in 'Materials and Methods'. The assay was prepared in V-bottom microtiter plates using the indicated effector:target ratios. These results are representative of those obtained in three similar experiments. Spontaneous lysis in this experiment was less than 10% for all time points.

and Methods'. A total of 5×10^5 CD4-depleted, transgenic splenocytes was used in order to provide 1×10^5 CD8+ T cells, based on FACS analysis. This number of CTL clone L3 cells was previously found to provide a strong signal in this assay. The assay plates were incubated until the lysate from the L3 CTL clone reached an absorbance level of 1.00. However, the lysate obtained from the CD4-depleted, transgenic splenocytes did not yield activity above background levels (<0.005 absorbance units at 405 nm; data not shown).

Kinetics of Proliferation by CD4-Depleted, Transgenic Splenocytes Stimulated with Allogenic Spleen Cells

Naive, CD4-depleted, transgenic splenocytes were stimulated with L^d alloantigen provided by T-depleted, DBA/2 splenocytes as described in 'Materials and Methods'. These cultures were then examined for proliferation and expression of cytolytic activity. The proliferative response was measured by the incorporation of ^{3}H-thymidine at day 1 through 7 of culture (fig. 2a). Proliferation of the naive, CD4-depleted, transgenic splenocytes stimulated with allogeneic spleen cells was not appreciable until day 2. Peak proliferative response occurred on day 5 with a pronounced decrease on day 6. Unstimulated cells did not proliferate.

Kinetics of Development of Cytolytic Activity by the Naive, CD4-Depleted, Transgenic Splenocytes Stimulated with Allogeneic Spleen Cells

Naive, CD4-depleted, transgenic splenocytes were cultured with T-depleted, DBA/2 splenocytes as described in 'Materials and Methods'. The responder population was collected daily for 5 days, and tested in an antigen-specific ^{51}Cr-release assay employing P815 mastocytoma target cells which express the L^d alloantigen (fig. 2b) and third-party EL-4 (H-2^b) target cells which do not express this antigen. The development of cytolytic activity closely paralleled the proliferative response. Only minimal antigen-specific cytolytic activity was evident after culture for 1 day. By the 2nd day of culture, cytolytic capacity had increased. Cytolytic activity increased further and reached maximal levels on day 3. Cytolytic response remained at maximal levels over the next 2 days, with 50% specific lysis observed at an effector:target cell ratio of less than 1:1 (fig. 2b). This cytolytic activity was antigen-specific as evidenced by the inhibition of cytolysis observed in the presence of anti-L^d mAb and by the failure of cultured cells to lyse a third-party target cell, EL-4 (data not shown). Cytolysis also was inhibited by anti-CD8 and anti-LFA-1 mAb, but not by anti-CD4 mAb (data not shown). For convenience in the following experiments, cells were obtained after culture of naive, CD4-depleted, transgenic splenocytes with T-depleted, DBA/2 splenocytes and 10 U/ml IL-2 for 3 days.

Expression of BLT Esterase Activity by the Naive, CD4-Depleted, Transgenic Splenocytes Stimulated with Allogeneic Spleen Cells Alone or Together with IL-2 or IL-4

Whole cell lysates were prepared from 1×10^5 CD4-depleted, transgenic splenocytes after 1 week in culture with alloantigen alone or together with IL-2 or IL-4. Lysates were also prepared from 1×10^5 CTL clone L3 cells obtained from maintenance cultures. Both sets of lysates were assayed for the presence of BLT esterase activity as described in 'Materials and Methods'. The assay plates were incubated until the lysate from the L3 CTL clone reached an absorbance level of 1.00. In contrast to results observed with freshly isolated CD4-depleted, transgenic splenocytes, CD4-depleted, transgenic splenocytes stimulated for 1 week in culture with alloantigen expressed significant BLT esterase activity. Lysates from such cells cultured with alloantigen alone produced absorbance levels ranging up to 0.5; lysates of cells from cultures also containing IL-2 produced absorbance levels ranging from 0.3 to 0.8, and lysates of cells from cultures also containing IL-4 produced absorbance levels in excess of 1.0 (data not shown).

Proliferation in Association with Stimulation of the TCR Generates Antigen-Specific Cytolytic Activity

The above results indicated that naive, CD4-depleted, transgenic splenocytes could be induced to proliferate and develop cytolytic activity following stimulation with L^d alloantigen in the form of T-depleted DBA/2 splenocytes. However, T-depleted DBA/2 splenocytes contain several different kinds of cells which express the relevant L^d alloantigen. P815 cells express L^d as evidenced by the fact that activated CD4-depleted,transgenic splenocytes were able to lyse these cells. Cul-

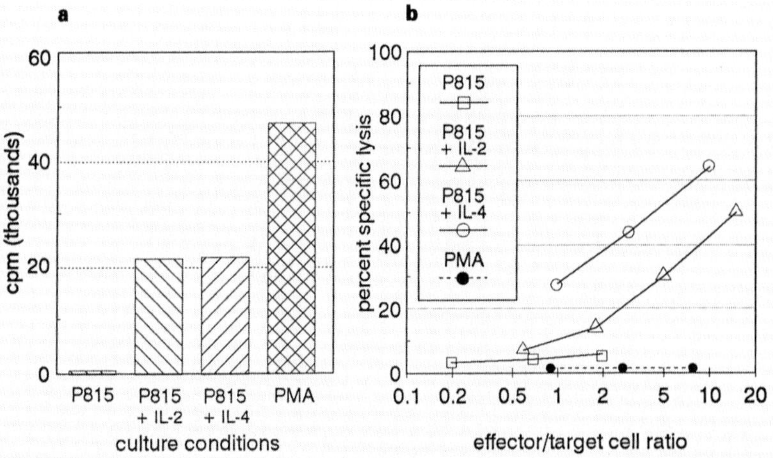

Fig. 3. a Proliferative response of naive, CD4-depleted, transgenic splenocytes stimulated with alloantigen-bearing P815 tumor cells, alone or together with IL-2 or IL-4, or to stimulation with PMA. Naive, CD4-depleted transgenic splenocytes (3×10^5) were cultured in triplicate with mitomycin-C-treated P815 tumor cells (3×10^5), alone or together with IL-2 (10 U/ml) or IL-4 (1,000 U/ml), in U-bottom microtiter plates or with PMA (50 ng/ml) as described in 'Materials and Methods'. Cultures were pulsed with ^{3}H-thymidine (1 μCi/well) for the final 12 h of a 72-hour culture; wells were harvested, and incorporated radioactivity was expressed as average counts per minute. These results are representative of those obtained in three similar experiments. **b** Acquisition of cytolytic activity by naive, CD4-depleted, transgenic splenocytes stimulated with alloantigen-bearing P815 tumor cells, alone or together with IL-2 or IL-4, or to stimulation with PMA. Cells were obtained from cultures prepared as described in the legend for figure 3a after 3 days and incubated with ^{51}Cr-labeled P815 target cells (2.5×10^3) in a standard 5-hour cytolytic assay as described in 'Materials and Methods'. Effector:target ratios for the different cell populations are indicated. Spontaneous ^{51}Cr-release was 7%. These results are representative of those obtained in three similar experiments.

tures were prepared using P815 mastocytoma cells as a homogeneous stimulating cell population to evaluate the relationship between TCR stimulation, proliferation, and the development of cytolytic capacity in a primary response. Naive, CD4-depleted, transgenic splenocytes were cultured with P815 mastocytoma cells alone or together with IL-2 or IL-4. Naive, CD4-depleted, transgenic splenocytes were also cultured in the presence of the phorbol ester PMA, since previous studies had shown that PMA alone was sufficient to induce proliferation of CD8+ CTL clones [36]. Cultures were assessed at day 3 for proliferation of CD8+ CTL clones [36]. Cultures were assessed at day 3 for proliferative activity as measured by incorporation with ^{3}H-thymidine and for cytolytic activity with a ^{51}Cr-release assay using P815 mastocytoma target cells.

Naive, CD4-depleted, transgenic splenocytes stimulated with P815 alone as the source of alloantigen did not proliferate (fig. 3a) or acquire cytolytic activity (fig. 3b). However, when IL-2 or IL-4 was included in addition to P815 mastocytoma cells, CD4-depleted, transgenic splenocytes both proliferated (fig. 3a) and developed cytolytic activity (fig. 3b). Stimulation with PMA alone induced a proliferative response, in some in-

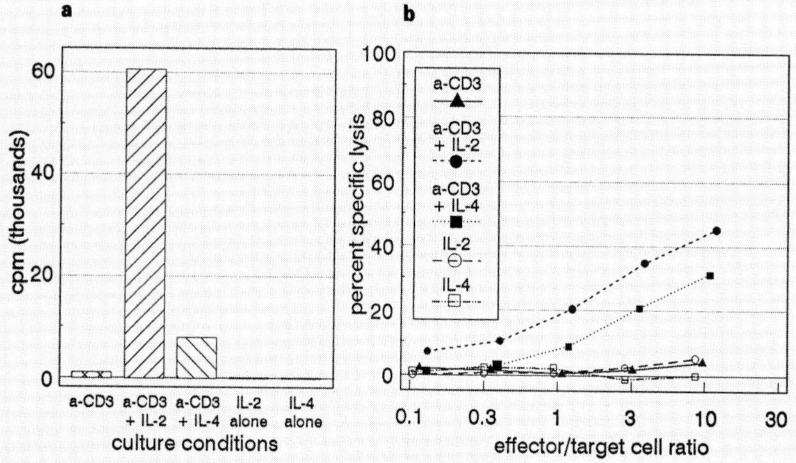

Fig. 4. a Proliferative response of purified, naive CD8+ transgenic T cells stimulated with IL-2, IL-4, immobilized anti-CD3ε mAb, either alone or together. Purified, naive, CD8+, transgenic T cells (2 × 10^4) were cultured in triplicate with immobilized anti-CD3 mAb (1:10 dilution of hybridoma culture supernatant), either alone or together with IL-2 (15 U/ml) or IL-4 (1,000 U/ml) pulsed with ^{3}H-thymidine (1 μCi/well) for the final 12 of 72 h of culture and harvested. The incorporated radioactivity was expressed as the average counts per minute. ^{3}H-thymidine incorporation by unstimulated purified, naive CD8+ transgenic T cells was 62 cpm. These data are representative of results obtained in two experiments. **b** Acquisition of cytolytic activity by purified, naive CD8+ transgenic T cells stimulated with immobilized anti-CD3ε mAb, either alone or together with IL-2 or IL-4. Cells were obtained after 3 days from cultures prepared as described in the legend for figure 4a and incubated with ^{51}Cr-labeled P815 target cells (2.5 × 10^3) in a standard 5-hour cytolytic assay as described in 'Materials and Methods'. Effector:target ratios for the different cell populations are indicated. Spontaneous ^{51}Cr-release was less than 7%. These data were obtained in a single experiment.

stances greater than that seen with cultures containing alloantigen and IL-2 or IL-4 (fig. 3a). However, this antigen-independent proliferation induced by PMA stimulation was not associated with the expression of cytolytic capacity (fig. 3b).

To simplify the model system further, naive CD8+ transgenic T cells were purified from the CD4-depleted splenocyte population by positive selection using a MACS column as described in 'Materials and Methods'. This population consisted of greater than 96% CD8+, transgene+ T cells as demonstrated by FACS analysis (data not shown). These purified CD8+ T cells were stimulated with immobilized anti-CD3ε mAb as described in 'Materials and Methods'. Following stimulation with anti-CD3ε mAb alone, the CD8+ transgenic T cells did not proliferate (fig. 4a) and did not acquire cytolytic activity (fig. 4b). In the experiment presented in figure 4, purified CD8+ transgenic T cells cultured with either IL-2 or IL-4 alone did not proliferate (fig. 4a) or become cytolytic (fig. 4b). However, purified CD8+ transgenic T cells stimulated with immobilized anti-CD3ε mAb in the presence of IL-2 both proliferated and developed cytolytic activity (fig. 4a, b). Purified CD8+ transgenic T cells stimulated with immobilized anti-CD3ε mAb in the presence of IL-4 did not proliferate or develop cytolytic activity as well as cells cultured with IL-2.

Thus, stimulation of the TCR of CD8+ TCR transgenic splenocytes with antigen or anti-CD3ε mAb alone does not induce proliferation or cytolytic activity. However, such stimulation does induce responsiveness to IL-2, and to a lesser extent to IL-4. Antigen-stimulated or anti-CD3ε mAb-stimulated CD4-depleted, TCR transgenic splenocytes or purified CD8+ transgenic T cells proliferate and acquire cytolytic activity following stimulation of the TCR if IL-2 is also present. Although proliferation of naive CD4-depleted, transgenic splenocytes appears to be necessary for the development of cytolytic activity, proliferation induced by PMA alone was not associated with acquisition of cytolytic activity. To evaluate further the relationship between proliferation and the acquisition of cytolytic activity, experiments were carried out using several pharmacological inhibitors of cell replication.

Proliferation of Naive, CD4-Depleted, Transgenic Splenocytes following Stimulation with Alloantigen and IL-2 Is Effectively Prevented by Chemicals Which Inhibit Cell Cycle Progression

Cultures were prepared by stimulating naive, CD4-depleted, transgenic spleoncytes or CTL clone L3 cells with irradiated, T-depleted DBA/2 splenocytes and IL-2, alone or with various pharmacological inhibitors of cell replication, for 3 days as described in 'Materials and Methods'. All cultures were pulsed for the final 12 h of a 72-hour culture with ^{3}H-thymidine, cultures were harvested, and the radioactivity incorporated by cells was measured and expressed as counts per minute. These culture conditions resulted in a good proliferative response as measured at day 3 in cultures prepared without the inhibitors (fig. 5a, b).

Progression through the G1/S phase of the cell cycle was inhibited by three agents which act through different mechanisms and at slightly different sites. Two were inhibitors of enzymes required for DNA synthesis: (1) aphidicolin, which reversibly inhibits DNA α-polymerase [37], and (2) hydroxyurea, which reversibly inhibits ribonucleotide diphosphoreductase [38]. The third G1/S phase inhibitor was mimosine, which arrests the cell cycle in late G1, at a stage just prior to the arrest point induced by aphidicolin [39]. The exact mechanism of action of mimosine is not known. However, it is an iron chelator [40, 41] and may inhibit an iron-dependent pathway in late G1 phase. Two agents were used to inhibit the cell cycle at the M phase: (1) colcemide, which is the classic inhibitor of microtubule function [42], and (2) nocodazole, an agent that acts on tubulin to inhibit microtubule formation by a different mechanism than colchicine [43]. Multiple agents acting at different phases of the cell cycle were used to lessen the possibility that the observed effects are uniquely attributable to a particular agent.

Incorporation of ^{3}H-thymidine by naive, CD4-depleted, transgenic splenocytes and by CTL clone L3 was inhibited by at least 99% in cultures containing the G1/S phase inhibitors and by at least 90% in cultures containing the M phase inhibitors, compared to cultures to which inhibitors were not added (fig. 5a, b). The low level of thymidine incorporation seen in cultures containing the M phase inhibitors may be due to a small percentage of responder cells that were entering the S phase at the time of pulsing with the ^{3}H-thymidine.

Acquisition of Cytolytic Activity but Not Lymphokine Production by Naive, CD4-Depleted, Transgenic Splenocytes Requires Proliferation in Response to Antigenic-TCR and IL-2 Stimulation

It has previously been reported that the cytolytic activity of CD8+ CTL clones does not vary with the phase of the cell cycle nor is

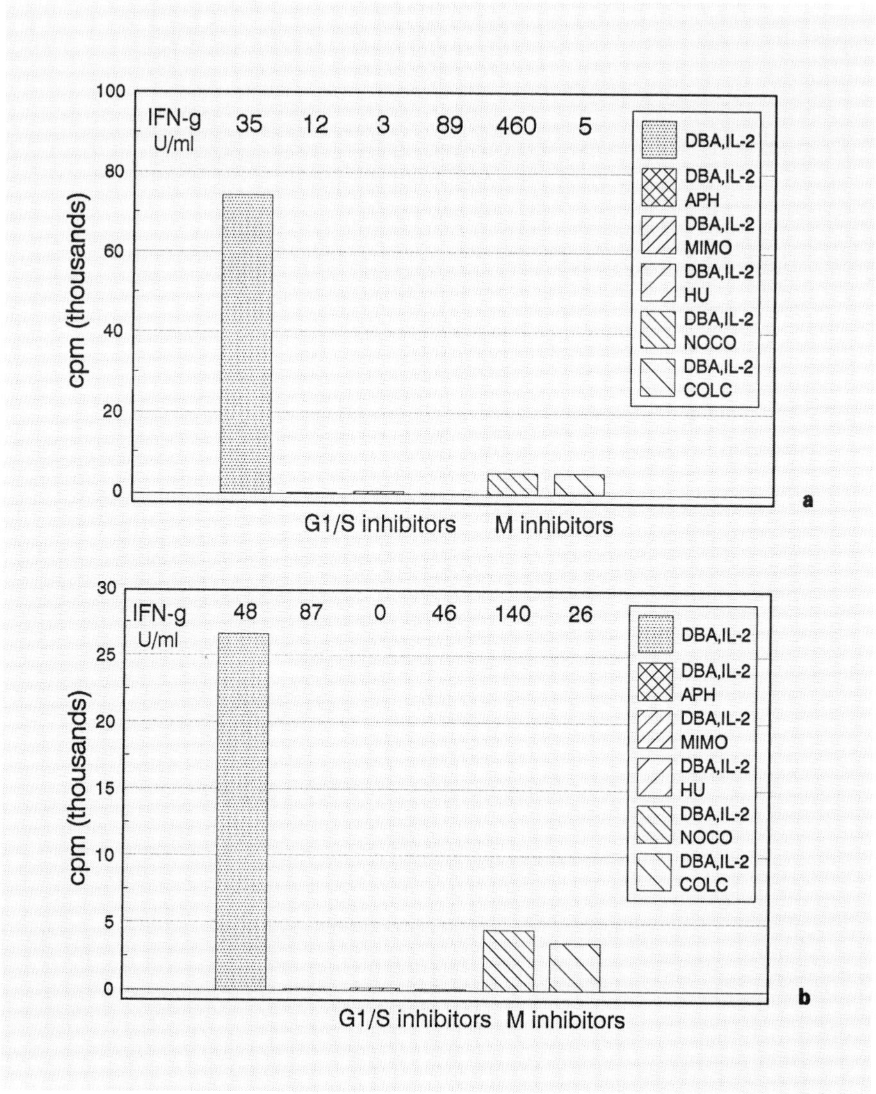

Fig. 5. a The proliferative response of CTL clone L3 cells in cultures containing allogeneic splenocytes and IL-2 is prevented by cell cycle inhibitors. CTL clone L3 cells (1×10^6) were collected from day 7 maintenance cultures, washed and incubated with T-depleted DBA/2 splenocytes (1×10^7) and IL-2 (10 U/ml) in 24-well plates in a final volume of 1.5 ml as described in 'Materials and Methods'. Replicate cultures were prepared, containing one of the following cell cycle inhibitors at the indicated final concentration. G1/S inhibitors: aphidicolin (APH, 5 μg/ml), mimosine (MIMO, 200 μ*M*), and hydroxyurea (HU, 1 m*M*); or M phase inhibitors nocadazole (NOCO, 0.04 μg/ml), and colcemide (COLC, 0.03 μg/ml) were added at the initiation of the cultures. After culture for 60 h, 200-μl aliquots of the cultures were transferred in triplicate to U-bottom microtiter plates and pulsed with ^{3}H-thymidine (1 μCi/well) for an additional 12 h. Wells were harvested and radioactivity expressed as average counts per minute. These results are representative of those obtained in two experiments. To measure production of IFN-γ, supernatants were collected from individual macrowells after 60 h of culture. All samples were stored at –20 °C until tested for IFN-γ activity by the ELISA method as described in 'Materials and Methods'. The lower limit of detection in this

their cytolytic activity affected by cell cycle inhibitors [44]. The relationship between proliferation and cytolytic capacity was compared in naive, CD4-depleted transgenic lymphocytes and the conventional CD8+ CTL clone L3 collected from the cultures described above. The cytolytic activity of these 3-day cultures was assayed using a 5-hour ^{51}Cr-release assay and P815 mastocytoma as target cells. Stimulation with alloantigen and IL-2 in the presence of the cell cycle inhibitors did not affect the observed cytolytic activity of CTL clone L3 cells (fig. 6a). The cytolytic activity of L3 cells was the same whether or not cultured with the cell cycle inhibitors.

However, naive, CD4-depleted transgenic splenocytes stimulated with alloantigen and IL-2 and cell cycle inhibitors failed to develop cytolytic activity, compared to control cultures containing no inhibitor (fig. 6b). Similar effects were observed with inhibitors of the G1/S phase and M phase of the cell cycle. The failure to generate cytolytic activity did not reflect a general failure to activate T cells. Supernatants from the cell cultures used for determination of cytolytic activity were assayed for the presence of IFN-γ using an ELISA as described in 'Materials and Methods'. IFN-γ was detected under most culture conditions tested (fig. 5a, b). IFN-γ production by both CD4-depleted transgenic lymphocytes and the conventional CD8+ CTL clone L3 was inhibited by mimosine and colcemide, augmented by nocodazole, and unaffected by the other pharmacologic agents. The reasons for these differences are not known.

Thus, the development of cytolytic activity by naive CD4-depleted, transgenic splenocytes stimulated with alloantigen and IL-2 is prevented by inhibitors of cell proliferation acting at different stages of the cell cycle while cytolytic activity of an established CD8+ CTL clone is not affected by these inhibitors. This failure of antigen-stimulated CD4-depleted, transgenic splenocytes to develop cytolytic activity is not a reflection of general failure of T cell activation since the stimulated transgenic splenocytes were able to secrete IFN-γ. As shown above, transgenic splenocytes stimulated with PMA, which proliferated but failed to acquire cytolytic activity (fig. 3), did not secrete detectable IL-2, IL-4, or IFN-γ (data not shown). This observation further illustrates the dissociation between proliferation, cytolytic activity, and cytokine production.

assay is <0.5 U/ml. The results are representative of those obtained in two experiments. **b** The proliferative response of naive, CD4-depleted, transgenic splenocytes in cultures containing allogeneic splenocytes and IL-2 is prevented by cell cycle inhibitors. Freshly prepared naive, CD4-depleted, transgenic splenocytes (2×10^6) were incubated with T-depleted DBA/2 splenocytes (1×10^7) and IL-2 (10 U/ml) in 24-well plates in a final volume of 1.5 ml; replicate cultures were prepared, containing the cell cycle inhibitors, and cells transferred to measure incorporation of ^3H-thymidine as described in the legend for figure 5a. These results are representative of results obtained in five experiments with aphidicolin, three experiments with mimosine, nocodazole, and colcemide, and two experiments with hydroxyurea. To measure production of IFN-γ, supernatants were collected from individual macrowells after 60 h of culture. All samples were stored at −20 °C until tested for IFN-γ activity by the ELISA method as described in 'Materials and Methods'. The lower limit of detection in this assay is <0.5 U/ml. The results are representative of those obtained in two experiments.

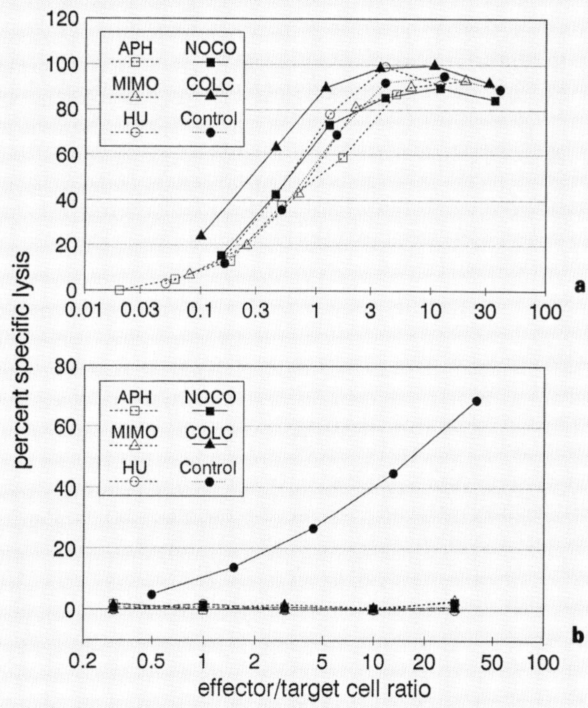

Fig. 6. a The cytolytic capacity of CTL clone L3 cells is unaffected by G1/S or by M phase cell cycle inhibitors. A standard 5-hour ^{51}Cr-release cytolytic assay was performed in V-bottom microtiter plates as described in 'Materials and Methods', using cells from the cultures described in the legend for figure 5A. Spontaneous ^{51}Cr release was less than 9%. These results are representative of those obtained in two experiments. **b** The development of cytolytic capacity by naive, CD4-depleted transgenic splenocytes stimulated with alloantigen and IL-2 in primary cultures is prevented by G1/S and by M phase cell cycle inhibitors. A standard 5-hour ^{51}Cr-release cytolytic assay was performed in V-bottom microtiter plates as described in 'Materials and Methods', using cells from the cultures described in the legend for figure 5b. Spontaneous ^{51}Cr release was less than 7%. These results are representative of those obtained in two experiments.

Failure of Naive, CD4-Depleted Splenocytes to Develop Cytolytic Activity Is Not Due to Toxicity of Cell Cycle Inhibitors

Cells from the cultures containing cell cycle inhibitors were compared to the control cultures in terms of cell viability and cell surface phenotype by FACS analysis (fig. 7). Viable CD8+ T cells, all of which expressed the transgene TCR, were present in similar proportions in control cultures and in cultures containing cell cycle inhibitors. Therefore, the failure to develop cytolytic activity in cultures containing the cell cycle inhibitors was not accounted for by a lack of CD8+, transgene-positive cells in the responding population.

Discussion

CD8+ T cells from mice which express an α/β transgenic TCR specific for L^d class I MHC antigen provide a model system which has many advantages for analysis of the requirements for the acquisition of cytolytic activity by naive T cells. All CD8+ T cells from these mice bear the same TCR [14]. Such T cells do not express cytolytic activity when freshly isolated [14] (and results presented above). We also found that such naive CD8+ transgenic T cells did not contain BLT esterase, an activity related to the serine esterase granzyme A which is located along with perforin in an intracellular granulosome [45, 46]. The failure to detect BLT esterase activity in

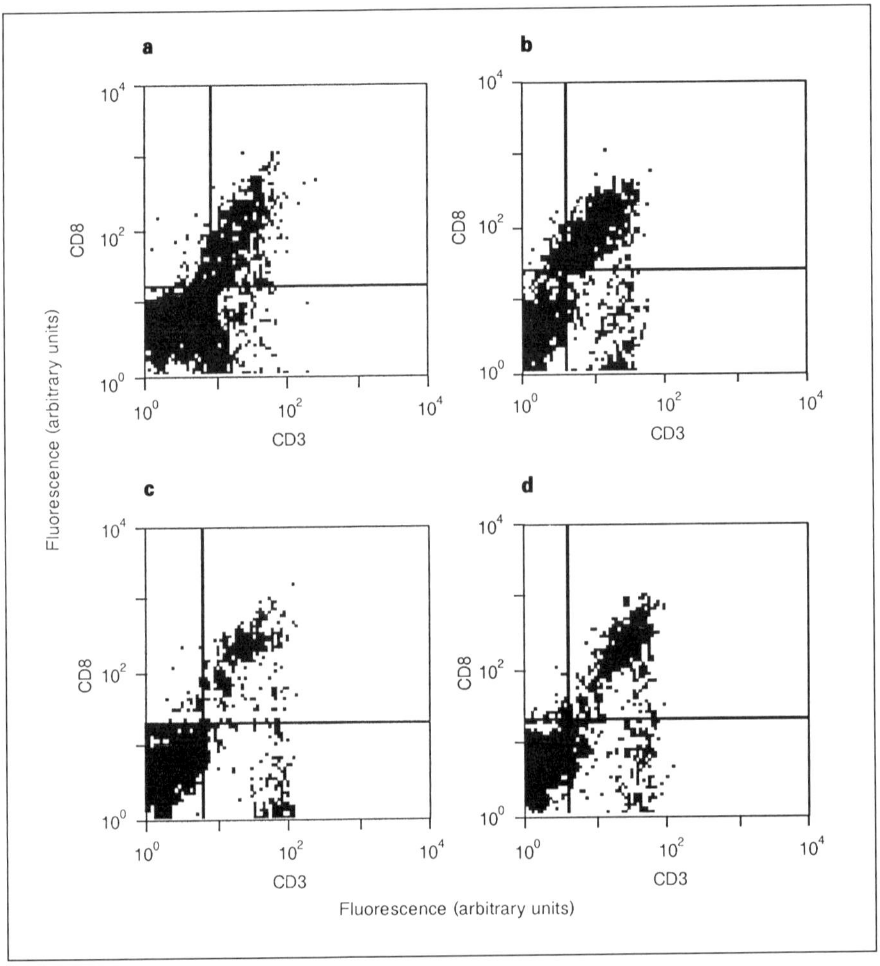

Fig. 7. Equivalent numbers of viable, CD8+, transgenic TCR+ T cells were collected after 3 days in culture prepared with or without cell cycle inhibitors. Cells were cultured as described in the legend for figure 5b, collected, washed, and centrifuged over a Ficoll-Hypaque density gradient. The resulting cell suspension was stained as described in 'Materials and Methods'. Two-color flow cytometry was performed using anti-CD8-PE (15 µl of 1:30 dilution) and anti-CD3-FITC (15 µl of 1:40 dilution). Propidium iodide was added to all samples, and viable cells were gated for analysis. Results are shown as dotplots of CD8 vs. CD3 and are from a single experiment which is representative of two similar experiments with cells staining with propidium iodide excluded. Figures represent CD8+ cells that express TCR (upper right quadrant). **a** Control cultures without inhibitor, 20%. **b** Culture containing the G1/S phase inhibitor mimosine, 22%. **c** Culture containing the G1/S phase inhibitor aphidicolin, 17%. **d** Culture containing the G2/M phase inhibitor nocodazole 24%. Cells stained with anti-CD3+ cells only (lower right quadrant) are a CD4–/CD8–, transgene+ population and represented from 14 to 15% of the cells in the various populations. This FACS result is representative of two similar experiments.

naive CD8+ transgenic T cells suggests that these cells do not express constitutively the mature phenotype of the granule-associated cytolytic pathway. Thus, the CD8+ T cells in the naive, CD4-depleted, transgenic splenocyte population are not inherently cytolytic.

When stimulated in culture with allogeneic, T-depleted splenocytes bearing the L^d alloantigen, these CD8+ transgenic T cells proliferate, secrete lymphokines, and develop antigen-specific cytolytic activity as well as BLT esterase activity. However, our data suggest that the kinetics of expression of these different parameters of T cell activation may not be the same. Lymphokine production begins soon after stimulation, and secretion of IFN-γ, a lymphokine that appears not to be utilized by the cultured cells, reaches maximal levels by 24 h following TCR stimulation (data not shown). In contrast, cytolytic activity was barely detectable at 24 h and did not reach maximal levels until 72 h.

Proliferation observed in cultures prepared using allogeneic, T-depleted splenocytes as stimulating cells appeared to be due largely to IL-2. IL-2 was detected in culture supernatants, and mAb reactive with IL-2 and with the IL-2 receptor inhibited proliferation in these cultures, but IL-4 production was not detected (data not shown). Thus, CD8+ transgenic T cells in cultures prepared using allogeneic T-depleted splenocytes were stimulated not only by antigen but by various costimulatory molecules expressed by the allogeneic cells and by IL-2 and other lymphokines secreted by the activated transgenic T cells. In an attempt to determine the relative contributions of TCR stimulation and proliferation to the generation of cytolytic activity, naive CD4-depleted transgenic splenocytes were stimulated through the TCR in the absence of costimulation. Two stimulating conditions were used: P815 mastocytoma cells which express the relevant L^d alloantigen but do not express B7-1 or B7-2, costimulatory molecules which can be involved in IL-2 secretion (data not shown), or immobilized anti-CD3ε mAb. Naive, CD4-depleted transgenic splenocytes stimulated with either P815 mastocytoma cells or immobilized anti-CD3ε mAb did not secrete either IFN-γ or IL-2 (data not shown), proliferation did not occur, and cytolytic activity did not develop.

However, TCR stimulation alone, using either P815 mastocytoma cells as a source of alloantigen or immobilized anti-CD3ε mAb, did induce responsiveness to IL-2 and to IL-4. Stimulation with either P815 mastocytoma cells or immobilized anti-CD3ε mAb in the presence of IL-2 or IL-4 led to proliferation and the expression of cytolytic activity. Neither IL-2 nor IL-4 alone induced proliferation of purified naive CD8+ transgenic T cells (fig. 4a). However, in some experiments IL-2 alone induced proliferation to levels up to one third that observed in the presence of anti-CD3 mAb or mastocytoma cells plus IL-2 when whole CD4-depleted transgenic spleen cells were used (data not shown). These cells expressed low levels of cytolytic activity at 24 h, but this activity had disappeared after 3 days of culture (data not shown). Activation of natural killer cell precursors, present in CD4-depleted transgenic spleen cells but probably missing in purified CD8+ populations, may account for these observations. IL-4 alone never induced either proliferation or expression of cytolytic activity.

Antigen-independent proliferation, induced by treatment of naive, CD4-depleted, transgenic splenocytes with the phorbol ester PMA, did not result in the development of antigen-specific cytolytic capacity. The ability of PMA alone to stimulate proliferation of CD8+ T cells was not unexpected since PMA had been shown to induce cytokine-independent proliferation of CD8+ murine CTL clones [36]. As was the case with CD8+ CTL clones [36], PMA did not induce production of IL-2, IL-4, or IFN-γ by naive, CD4-depleted transgenic splenocytes (data not shown). These observations suggest that proliferation alone is not sufficient to lead to the development of cytolytic ability. However, these results must be interpreted with caution since prolonged exposure to PMA has been found to abolish the cytolytic activity of cloned CTL [47, 48]. Thus, although the CD8+ T cells indeed proliferate in response to PMA in the absence of stimulation of the TCR, their failure to exhibit cytolytic activity may reflect an inactivation of the cytolytic mechanism through the prolonged stimulation of protein kinase C by PMA rather than their failure to develop cytolytic activity.

These results suggest that neither stimulation of the TCR nor cell proliferation alone is sufficient to lead to the efficient induction of cytolytic activity. To evaluate further the relationship between cell replication and the development of cytolytic activity, the effects of several cell cycle inhibitors that inhibited different key points of the cell cycle was measured in cultures prepared with alloantigen and exogenous IL-2. Multiple agents inhibiting the progression from the G1 into the S phase or blocking at the G2/M phase were employed to minimize agent-specific effects, and effects on CD4-depleted transgenic splenocytes and a cloned CD8+ T cell clone were compared. The effects of these inhibitors on proliferation of these two populations of cells were similar. G1/S phase inhibitors reduced incorporation of ^{3}H-thymidine approximately by 99%. M phase inhibitors reduced ^{3}H-thymidine incorporation to 90% of control levels; cells that were not inhibited probably were passing through the S phase during the time of exposure to the labeled thymidine.

However, the effects of cell cycle inhibitors on cytolytic activity of the two cell populations was strikingly different. The inhibitors completely prevented the development of cytolytic activity by naive, CD4-depleted transgenic splenocytes stimulated with irradiated, T-depleted allogeneic splenocytes and IL-2. In contrast, none of the inhibitors had any effect on the level of cytolytic activity expressed by the CD8+ CTL clone L3. The failure of antigen-stimulated CD4-depleted transgenic splenocytes to develop cytolytic activity in the presence of mitotic inhibitors was not due to selective death of CD8+ transgenic T cells. FACS analysis showed that there were similar proportions of viable, transgenic, CD8+ T cells in cultures with or without cell cycle inhibitors. The ability of antigen-stimulated CD4-depleted transgenic splenocytes to secrete IFN-γ in the presence of cell cycle inhibitors indicated that all manifestations of T cell activation were not impaired.

Collectively, these results indicate that stimulation of the TCR of naive, CD8+ transgenic T cells with T-depleted allogeneic spleen cells bearing the appropriate L^d alloantigen and expressing costimulatory molecules leads to full activation. Cells proliferate, produce lymphokines, and acquire cytolytic activity. In contrast, stimulation of the TCR of naive, CD8+ transgenic T cells in the absence of relevant costimulatory molecules, either with immobilized anti-CD3ε mAb or with P815 mastocytoma cells that express L^d alloantigen but not the appropriate costimulatory molecules, does not induce any of these events. However, stimulation of CD8+ transgenic T cells with antigen in the absence of relevant costimulatory molecules does render cells sensitive to the effects of IL-2 and IL-4. Proliferation of the naive, CD8+ transgenic T cells in response to IL-2 or IL-4 appears to be necessary for the development of cytolytic activity since pharmacologic agents that inhibit cell proliferation, either at the G1/S or M phases of the cell cycle, prevent the acquisition of cytolytic activity but do not affect production of the lymphokine IFN-γ. Proliferation may not be sufficient for the development of cytolytic capability since PMA induces a high level of proliferation of naive, CD8+ transgenic T cells, but cells do not become cytolytic or produce lymphokines, although this finding must be considered in light of possible effects of PMA on the cytolytic process.

Thus, stimulation of the TCR of naive, CD8+ transgenic T cells appears to induce cytolytic activity only if cell proliferation occurs in response to T cell growth factors. Growth-promoting lymphokines can be secreted by stimulated CD8+ T cells if appropriate costimulation is provided along with antigen. Alternatively, IL-2 or IL-4 may be provided by stimulated CD4+ helper T cells or by exogenous sources.

Acknowledgments

This research was supported by Research Grants CA-44372, AI-29531, CA-19266, and the Cancer Center Support Grant CA-14599 from the US Public Health Service.

The authors wish to thank Dr. Dennis Loh for providing the α/β-TCR transgenic mice and the anticlonotypic mAb which reacts with this receptor, Dr. Jeffrey Bluestone for the anti-CD3ε mAb, Dr. William Paul for the anti-IL-4 mAb, Dr. Tim Mosmann for the anti-IL-2 mAb, and Dr. Robert Schreiber for the reagents for the IFN-γ ELISA assay. The authors wish to acknowledge the technical assistance of Yukio Hamada. Also, we wish to thank Mrs. Frances Mills for help in preparing the manuscript.

References

1 Häyry P, Andersson LC, Nordling S, Virolainen M: Allograft response in vitro. Transplant Rev 1972;12:91–140.
2 MacDonald HR, Phillips RA, Miller RG: Allograft immunity in the mouse. II. Physical studies of the development of cytotoxic effector cells from their immediate progenitors. J Immunol 1973;111:575–589.
3 Röllinghoff M, Schrader J, Wagner H: Effect of azathioprine and cytosine arabinoside on humoral and cellular immunity in vitro. Clin Exp Immunol 1973;15:261–269.
4 Cantor H, Jandinski J: The relationship of cell division to the generation of cytolytic activity in mixed lymphocyte culture. J Exp med 1974;140:1712–1716.
5 Clark W, Nedrud J: Effect of BUdR on proliferation and development of cytotoxicity in mixed leukocyte culture. Cell Immunol 1974;10:159–164.
6 Denizot F, Wilson A, Battye F, Berke G, Shortman K: Clonal expansion of T cells: A cytotoxic T-cell response in vivo that involves precursor cell proliferation. Proc Natl Acad Sci USA 1986;83:6089–6092.
7 Alter BJ, Grillot-Courvalin C, Bach ML, Zier KS, Sondel PM, Bach FH: Secondary cell-mediated lympholysis: Importance of H-2 LD and SD factors. J Exp Med 1976;143:1005–1014.
8 Nabholz M, MacDonald HR: Cytolytic T lymphocytes. Annu Rev Immunol 1983;1:273–306.
9 Trenn G, Takayama H, Hu-Li J, Paul WE, Sitkovsky MV: B cell stimulatory factor 1 (IL-4) enhances the development of cytotoxic T from Lyt-2^{+} resting murine T lymphocytes. J Immunol 1988;140:1101–1106.
10 Okada M, Kitahara M, Kishimoto S, Matsuda T, Hirano T, Kishimoto T: IL-6/BSF-2 functions as a killer helper factor in the in vitro induction of cytotoxic T cells. J Immunol 1988;141:1543–1549.
11 Chen W-F, Zlotnik A: IL-10: A novel cytotoxic T cell differentiation factor. J Immunol 1991;147:528–534.
12 Gromo G, Geller RL, Inverardi L, Bach FH: Signal requirements in the step-wise functional maturation of cytotoxic T lymphocytes. Nature 1987;327:424–426.
13 Wagner H, Hardt C, Rouse BT, Rollinghoff M, Scheurich P, Pfizenmaier K: Dissection of the proliferative and differentiative signals controlling murine cytotoxic T lymphocyte responses. J Exp Med 1982;155:1876–1881.
14 Sha WC, Nelson CA, Newberry RD, Kranz DM, Russell JH, Loh DY: Selective expression of an antigen receptor on CD8-bearing T lymphocytes in transgenic mice. Nature 1988;335:271–274.
15 Glasebrook AL, Fitch FW: Alloreactive cloned T cell lines. I. Interactions between cloned amplifier and cytolytic T cell lines. J Exp Med 1980;151:876–895.
16 Leo O, Foo M, Sachs DH, Samelson LE, Bluestone JA: Identification of a monoclonal antibody specific for a murine T3 polypeptide. Proc Natl Acad Sci USA 1987;84:1374–1378.
17 Heinzel FP, Sadick MD, Holaday BJ, Coffman RL, Locksley RM: Reciprocal expression of interferon γ or interleukin 4 during the resolution or progression of murine leishmaniasis: Evidence for expansion of distinct helper T cell subsets. J Exp Med 1989;169:59–72.
18 Cherwinski HM; Schumacher JH, Brown KD, Mosmann TR: Two types of mouse helper T cell clone. III. Further differences in lymphokine synthesis between Th1 and Th2 clones revealed by RNA hybridization, functionally monospecific bioassays, and monoclonal antibodies. J Exp Med 1987;166:1229–1244.
19 Kranz DM, Tonegawa S, Eisen HN: Attachment of an anti-receptor antibody to non-target cells renders them susceptible to lysis by a clone of cytotoxic T lymphocytes. Proc Natl Acad Sci USA 1984;81:7922–7926.

20 Spitalny GL, Havell EA: Monoclonal antibody to murine gamma interferon inhibits lymphokine-induced antiviral and macrophage tumoricidal activities. J Exp Med 1984;159:1560–1565.

21 Sarmiento M, Dialynas DP, Lancki DW, Wall KA, Lorber MI, Loken MR, Fitch FW: Cloned T lymphocytes and monoclonal antibodies as probes for cell surface molecules active in T cell-mediated cytolysis. Immunol Rev 1982;68:135–169.

22 Sarmiento M, Glasebrook AL, Fitch FW: IgG or IgM monoclonal antibodies reactive with different determinants on the molecular complex bearing Lyt 2 antigen block T cell-mediated cytolysis in the absence of complement. J Immunol 1980;125:2665–2672.

23 Dialynas DP, Quan ZS, Wall KA, Pierres A, Quintans J, Loken MR, Pierres M, Fitch FW: Characterization of the murine T cell surface molecule, designated L3T4 identified by monoclonal antibody GK1.5: Similarity of L3T4 to the human Leu-3/T4 molecule. J Immunol 1983;131:2445–2451.

24 Sarmiento M, Loken MR, Fitch FW: Structural differences in cell surface T25 polypeptides from thymocytes and cloned T cells. Hybridoma 1981;1:13–26.

25 Unkeless JC: Characterization of a monoclonal antibody directed against mouse macrophages and lymphocyte Fc receptors. J Exp Med 1979;150:580–596.

26 Lowenthal JW, Corthesy P, Tougne C, Lees R, MacDonald HR, Nabholz M: High and low affinity IL 2 receptors: Analysis by IL 2 dissociation rate and reactivity with monoclonal anti-receptor antibody PC61. J Immunol 1985;135:3988–3994.

27 Lancki DW, Ma DI, Havran WL, Fitch FW: Cell surface structures involved in T cell activation. Immunol Rev 1984;81:65–94.

28 Davidson WF, Parish CR: A procedure for removing red cells and dead cells from lymphoid cell suspensions. J Immunol Methods 1975;7:291–300.

29 Nau GJ, Moldwin RL, Lancki DW, Kim D-K, Fitch FW: Inhibition of IL 2-driven proliferation of murine T lymphocyte clones by supraoptimal levels of immobilized anti-T cell receptor monoclonal antibody. J Immunol 1987;139:114–122.

30 Cerottini J-C, Engers HD, MacDonald HR, Brunner KT: Generation of cytotoxic T lymphocytes in vitro. I. Response of normal and immune mouse spleen cells in mixed leukocyte cultures. J Exp Med 1974;140:703–717.

31 Coleman PL, Green GJ: A coupled photometric assay for plasminogen activator. Methods Enzymol 1981;80:408–414.

32 Schreiber RD, Hicks LJ, Celada A, Buchmeier NA, Gray PW: Monoclonal antibodies to murine γ-interferon which differentially modulate macrophage activation and antiviral activity. J Immunol 1985;134:1609–1618.

33 Schell SR, Nelson DJ, Fozzard HA; Fitch FW: The inhibitory effects of K^+ channel-blocking agents on T lymphocyte proliferation and lymphokine production are 'nonspecific'. J Immunol 1987;139:3224–3230.

34 Mosmann T: Rapid colorimetric assay for cellular growth and survival: Application to proliferation and cytotoxicity assays. J Immunol Methods 1983;65:55–63.

35 Malek TR, Ortega RG, Jakway JP, Chan C, Shevach EM: The murine IL 2 receptor. II. Monoclonal anti-IL 2 receptor antibodies as specific inhibitors of T cell function in vitro. J Immunol 1984;133:1976–1982.

36 Kim D-K, Otten G, Moldwin RL, Dunn DE, Nau GJ, Fitch FW: PMA alone induces proliferation of some murine T cell clones but not others. J Immmunol 1986;137:2755–2760.

37 Huberman JA: New views on the biochemistry of eucaryotic DNA replication revealed by aphidicolin, an unusual inhibitor of DNA polymerase α. Cell 1981;23:647–648.

38 Reichard P, Ehrenberg A: Ribonucleotide reductase – A radical enzyme. Science 1983;221:514–519.

39 Lalande M: A reversible arrest point in the late G1 phase of the mammalian cell cycle. Exp Cell Res 1990;186:332–339.

40 Kontoghiorghes GJ, Evans RW: Site specificity of iron removal from transferrin by alpha-ketohydroxypyridine chelators. FEBS Lett 1985;189:141–144.

41 Forsbek K, Nilsson K, Kontoghiorghes GJ: Variation in iron accumulation, transferrin membrane binding and DNA synthesis in the K-562 and U-937 cell lines induced by chelators and their iron complexes. Eur J Haematol 1987;39:318-325.

42 Salmon ED, McKeel M, Hays T: Rapid rate of tubulin dissociation from microtubules in the mitotic spindle in vivo measured by blocking polymerization with colchicine. J Cell Biol 1984;99:1066–1075.

43 DeBradaner M, Geuens G, Nuydens R, Willebrords R, Aerts F, DeMay J: Microtubule dynamics during cell cycle: The effects of taxol and nocodazole on the microtubule system of Ptk2 cells at different stages of the mitotic cycle. Int Rev Cytol 1986;101:215–274.

44 Sekaly RP, MacDonald HR, Zaech P, Glasebrook AL, Cerottini J-C: Cytolytic T lymphocyte functions is independent of growth phase and position in the mitotic cycle. J Exp Med 1981;154:575–580.

45 Podack ER, Hengartner H, Lichtenheld MG: A central role of perforin in cytolysis. Annu Rev Immunol 1991;9:129–157.

46 Peters PJ, Borst J, Oorschot V, Fukuda M, Krähenbühl O, Tschopp J, Slot JW, Geuze HJ: Cytotoxic T lymphocyte granules are secretory lysosomes, containing both perforin and granzymes. J Exp Med 1991;173:1099–1109.

47 Orosz CG, Roopenian DC: Phorbol myristate acetate and in vitro T lymphocyte function. III. Selective impairment by PMA of lethal hit delivery by cloned CTL. Transplantation 1985;39:411–418.

48 Sano S, Kiyotaki C, Tatsumi Y, Fujiwara H, Hamaoka T: Cytotoxic T lymphocyte unresponsiveness induced by prolonged treatment with immobilized anti-CD3 antibody: Association of impairment of cytolytic activity with temporary depletion of intracellular protein kinase C. J Immunol 1989;143:2797–2805.

Immunol Res 1994;13:234–243

Jonathan M. Green[a, d]
Craig B. Thompson[a–e]

[a] Department of Medicine,
[b] Department of Molecular Genetics and Cell Biology,
[c] Howard Hughes Medical Institute,
[d] Gwen Knapp Center for Lupus and Immunology Research,
[e] Committee on Immunology,
University of Chicago, Ill., USA

Modulation of T Cell Proliferative Responses by Accessory Cell Interactions

Key Words
T lymphocyte
Costimulation
CD28
LFA-1
Accessory cells

Abstract
Antigen-specific activation of the T cell is accomplished by engagement of the T cell receptor (TCR) by an antigen (Ag)/MHC complex presented on the surface of an antigen-presenting cell (APC). However, it has been demonstrated that engagement of the TCR by Ag/MHC complexes alone is normally insufficient to lead to a proliferative response and the development of effector function. Thus it has been proposed that the APC also provides additional signals which serve to modulate the T cell's response. These second or costimulatory signals are thought to be critical in the generation of a T cell-driven immune response. Several receptors have been proposed to be capable of serving as costimulatory receptors. Candidate molecules include CD28 and LFA-1 as well as other receptors. In this review the studies that we have performed to clarify the role of both LFA-1 and CD28 in providing costimulatory activity for T cell activation are discussed. In addition, we present evidence that under certain conditions, TCR signalling alone can be sufficient to lead to T cell proliferation.

Introduction

The accessory cell plays a multifactorial role in modulating the immune response. One critical function performed by the accessory cell is the processing and presentation of exogenous antigen (Ag) to the T lymphocyte. Exogenous Ag alone is not capable of binding to the T cell receptor (TCR) in a productive manner. It must first be taken up and processed by an antigen-presenting cell (APC). Antigen processing involves proteolysis and incorporation of peptide fragments into the Ag-binding groove of a class MHC molecule [for recent review, see ref. 1]. The Ag/MHC complexes are transported to the surface of the APC where they are available to interact with a T cell bearing a TCR that is restricted to the appropriate MHC haplotype and is specific for the particular antigen presented (fig. 1). If such an encounter occurs, productive engagement of the TCR may result, leading to T cell activation, clonal expansion and the development of effector function. Examples of cells that are capable of processing and presenting exogenous antigen include monocytes, B cells,

A portion of this work has previously been published in part in *Eur J Immunol* 1994;24:265–272.

Dr. Craig B. Thompson
The University of Chicago
Gwen Knapp Center
924 E. 57 Street, R413A
Chicago, IL 60637–5420 (USA)

0257–277X/94/
0134–0234$8.00/0

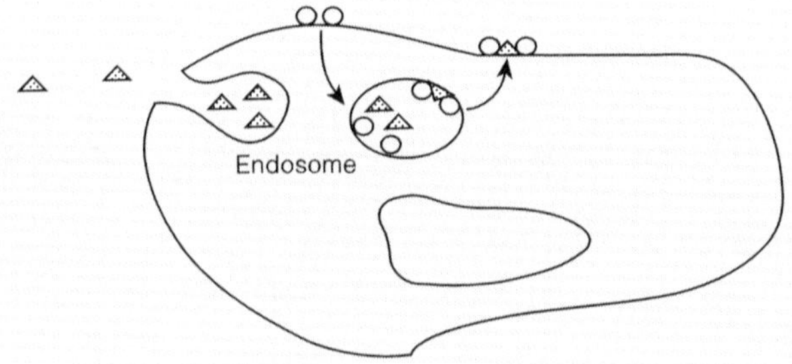

Fig. 1. Schematic depicting processing and presentation of exogenous antigen by an APC. The exogenous antigen is first internalized, and then incorporated into the peptide-binding groove of a class II MHC molecule, and the complex is then expressed on the surface of the APC.

and dendritic cells. Under certain conditions, other cell types may be induced to express class II MHC complexes and thus may play a role in antigen presentation.

While engagement of the TCR is a critical event in the initiation of a T cell response, it has been proposed that an additional signal or signals may also be required. The two-signal hypothesis for lymphocyte activation was first proposed by Bretscher and Cohn [2] as a model to explain self-versus-non-self discrimination in the generation of a humoral immune response. The model was subsequently applied to the T cell by Lafferty and Cunningham [3] in 1975. Since then considerable data has accumulated to support this hypothesis. Schwartz [4] and Quill and Schwartz [5] demonstrated that engagement of the TCR with Ag/MHC complexes in planar lipid membranes did not result in a T cell proliferative response. In fact, such TCR engagement led to a state of prolonged nonresponsiveness to further antigenic challenge. Further studies demonstrated that the required costimulatory activity could be provided in trans by the addition of APCs [6]. Intercellular contact was required for this activity and it was shown to be delivered via receptors on the surface of the T cell other than the antigen-specific TCR [7]. The requirement for signals in addition to TCR engagement to induce a proliferative response could provide a mechanism for preventing reactivity to self antigens not encountered in the thymus.

The identity of the molecule(s) capable of providing a costimulatory signal has been an area of intensive investigation. A number of candidates have been proposed, some soluble, such as interleukin-1, as well as many surface receptors such as LFA-1, CD2, HSA, CD44 and CD28. Each of these possess the ability to augment the T cells' proliferative response to antigenic challenge under certain circumstances. However, the mechanism by which each acts is probably distinct. An obstacle to achieving agreement on which T cell surface proteins function as costimulatory receptors lies in accepting a uniform definition of costimulation. In a practical sense, a costimulatory receptor must be able to induce the proliferation of a T cell activated with a submitogenic level of antigen, and function by a mechanism independent of the TCR.

A difficulty with rigorous adherence to the two-signal model for T cell activation is that it ignores the observation that cultures of highly purified T cells that have been depleted of any functional accessory cell activity can still be induced to proliferate by high doses of anti-CD3 antibodies [8, 9]. This would suggest that if in fact a sufficiently strong signal is provided through the TCR/CD3 complex alone, costimulation may not be required. In vivo, this situation may not occur except under circumstances of very high antigenic burden. In the more usual circumstance, each antigen is probably present at a relatively low concentration on the surface of the APC. In addition, the affinity of the TCR for the Ag/MHC complex is relatively low, with a K_d on the order of 10^{-4} to 10^{-5} M [10]. Thus, most of the time, an additional signal is mandatory. Simplistically, one could envision that this interaction may function in one of two ways. One mechanism might be to increase the avidity of the TCR-Ag/MHC interaction and thereby facilitate the cross-linking of multiple TCR/CD3 complexes, thus raising the strength of this interaction above the threshold needed for activation and proliferation. Alternatively, the second signal might function to lower the threshold needed for activation so that a lesser degree of TCR occupancy would result in a suprathreshold stimulus. Many of the adhesion receptors appear to be able to 'costimulate' by the former mechanism. An example of this, discussed below, is LFA-1. In contrast, costimulation via a signalling pathway that is independent of TCR signalling is a property of the CD28 receptor.

Enhancement of Proliferation through LFA-1 Is a Consequence of Enhancing Signal One

LFA-1 is a member of the integrin family. It is a heterodimeric transmembrane protein composed of an α_l (CD11a) and β_2 (CD18) subunit that are noncovalently associated. LFA-1 binds to at least three distinct ligands, ICAM-1, -2 and -3, that are present on a wide variety of cell types. Binding of LFA-1 to its ligands has been shown to augment the proliferation of antigen-activated T cells [11]. However, it has been unclear as to whether LFA-1 ligand interactions are truly costimulatory, or whether the enhanced intercellular adhesion increases the TCR binding to Ag/MHC complexes, effectively increasing the signal delivered through the TCR.

The regulatory properties of monoclonal antibodies against LFA-1 were initially described by van Noesel et al. [12] in 1988. They noted that antibodies directed against CD11a synergized with immobilized anti-CD3 in the induction of a proliferative response, whereas anti-CD18 antibodies tended to inhibit proliferation. Other groups have used purified ICAM proteins or ICAM-Ig fusion molecules and demonstrated that coimmobilization of anti-CD3 antibodies with ICAM augments the T cell proliferative response [13–16]. Whether engagement of LFA-1 in trans is capable of augmenting proliferation is less clear. Van Seventer et al. [17] were able to demonstrate costimulation in trans using purified ICAM-1 immobilized to beads [17]. However, expression of ICAM-1 on either CHO cells or L cells has been insufficient to provide costimulation in trans unless cotransfected with class II MHC [18, 19].

Superantigens have been shown to be able to bind to both the TCR and to class II MHC molecules [20]. Each is engaged by a different portion of the superantigen, and the binding is to regions of the TCR and class II complex not utilized in the binding of traditional antigens (fig. 2). Superantigens are monovalent with respect to binding the TCR, thus are not able to directly cross-link the TCR. In fact, engagement of the TCR by superantigens in the absence of accessory cells is insufficient for the induction of a proliferative response. This has been interpreted as an absolute requirement for the presence of class II, so that a three-way complex of SA/TCR/MHC can be formed. In this way, effective TCR cross-linking can occur, as well as enabling the provision of any other accessory cell-derived signals. Examples of superantigens include the staphylococcal enterotoxins, toxic shock syndrome toxin 1 as well as the Mls antigens [20–25].

In studying the costimulatory requirements for purified T cells activated with the bacterial toxins, staphylococcal enterotoxin A and B (SEA and SEB), we determined that in the absence of accessory cells, engagement of CD28 was sufficient to induce a proliferative response [26]. This response was dependent on LFA-1 (fig. 3a). In contrast, when the T cell is activated by a combination of phorbol ester plus CD28 ligation, blockade of LFA-1 has no effect, as the requirement for TCR engagement has been pharmacologically bypassed (fig. 3b). Thus, when T cells are activated by superantigen binding to the TCR, LFA-1-dependent adhesion may serve to stabilize the SA/TCR interactions, thus allowing TCR cross-linking to occur.

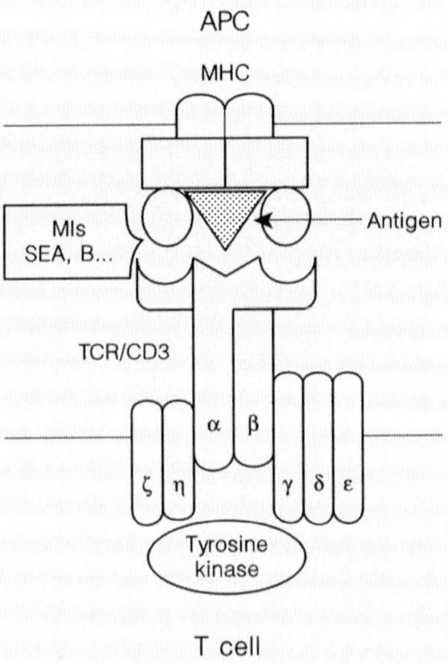

Fig. 2. Interaction of nominal antigen or superantigens with both the TCR and class II MHC complex.

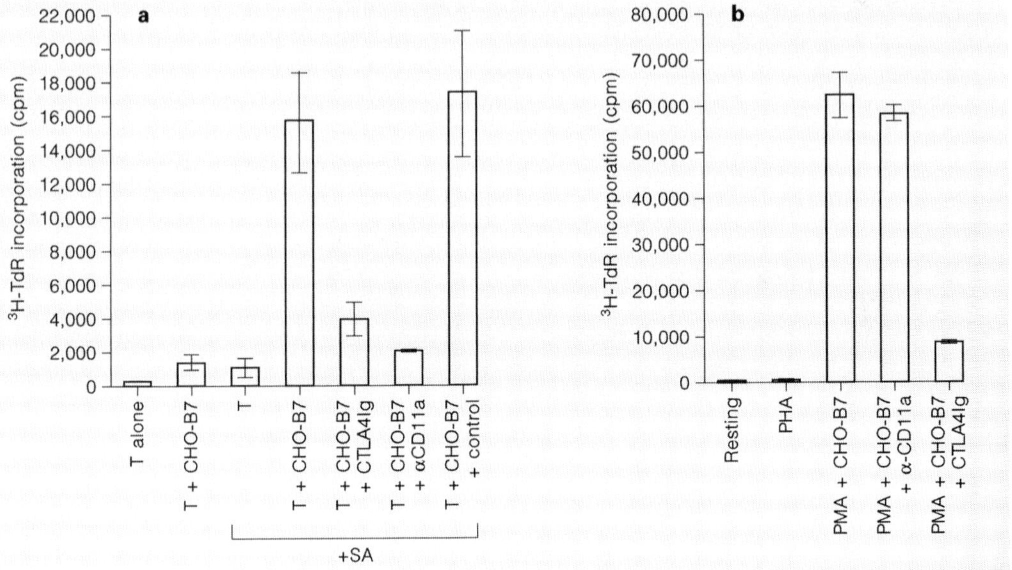

Fig. 3. Purified resting human T cells were activated by either a combination of SEA and SEB (**a**) or PMA (**b**) alone, or in the presence of B7-transfected CHO cells with CTLA4Ig, anti-CD11a, or a control immunoglobulin.

As evidence against a role for LFA-1 in providing an independent costimulatory signal, we have shown that engagement of LFA-1 with its natural ligand, ICAM-1 does not synergize with submitogenic TCR stimulation [18]. Thus LFA-1 plays a crucial role in stabilizing TCR interactions with an Ag/MHC complex, but is not in itself delivering an independent signal sufficient for the induction of proliferation. Disruption of LFA-1/ligand interactions prevents effective TCR cross-linking and blocks the induction of a proliferative response by preventing the induction of TCR-derived activation signals. In vivo, LFA-1 may function in an analogous manner. When the TCR is engaged by a low affinity ligand, such as most Ag/MHC complexes, LFA-1 may facilitate the TCR/Ag/MHC interaction, effectively increasing the strength of signal one. This is supported by the observation that ICAM-1 transfected cells can only augment the proliferative response of antigen-activated T cells if they coexpress class II MHC [11, 19]. Mechanistically, it has been shown that stimulation of T cells with coimmobilized ICAM-1 and anti-CD3 leads to a prolonged hydrolysis of inositol phospholipids and an increased accumulation of inositol phosphates as well as a sustained rise in intracellular calcium concentrations [27], essentially an enhancement of the intracellular events that occur upon TCR engagement.

Thus the bulk of the evidence favors a role for LFA-1 in promoting adhesion between the T cell and the APC, thus enhancing the signalling through the TCR. Whether it is a purely mechanical phenomenon by promoting intercellular contact and increased TCR cross-linking, or a more complex signal transduction event is less clear. In either case, costimulation through LFA-1 can raise a subthreshold stimulus to suprathreshold levels, resulting in T cell proliferation.

CD28 Provides a Novel Costimulatory Signal That Is Independent of TCR/CD3 Signalling

Engagement of CD28 was first shown to enhance the proliferation of human T cells activated with phytohemagglutinin by Gmünder and Lesslauer [28] in 1984. Since then numerous studies have confirmed the ability of CD28 ligation by either monoclonal antibody or natural ligand to induce a proliferative response of T cells activated by submitogenic doses of antigen or anti-CD3 antibodies [for recent review, see ref. 29]. A novel property of CD28 is its ability to induce a T cell proliferative response that is resistant to suppression by cyclosporin A [30]. The signal transduction pathway induced by CD28 ligation has not been well characterized, but appears to involve the induction of protein tyrosine kinase activity, and is clearly distinct from events induced through TCR stimulation [31].

The ligands for CD28 have been identified and characterized as the B7 family of proteins [32–35]. These are present on most professional APCs, as well as on some cells of epithelial origin [36, 37]. In addition to interacting with CD28, B7 molecules have been shown to bind to a second counterreceptor, CTLA-4 [38]. CTLA-4 is not expressed on resting T cells, but is induced upon activation. The affinity of B7 for CTLA-4 is higher than for CD28. This has been taken advantage of in the construction of a soluble chimeric molecule in which the extracellular portion of CTLA-4 is fused with the constant region of IgG_1. The fusion protein, CTLA4Ig, is a potent inhibitor of B7/ligand interactions. In vivo, treatment with CTLA4Ig leads to prolonged allograft and xenograft survival, as well as decreased T-dependent B cell responses [39–41]. In vitro, B7-dependent costimulation can be effectively inhibited by inclusion of CTLA4Ig in the cultures (fig. 3, 4).

The function of CTLA-4 has been debated amongst investigators. Damle et al. [42] have proposed that CTLA-4 functions as a costimulatory receptor in a manner analogous to CD28. Using partially activated CD4+ human T cells they demonstrated an augmentation of proliferation with anti-CTLA-4 antibodies. The increase in proliferation was less than that seen with CD28 costimulation. In contrast to CD28 ligation, they were not able to detect any increase in intracellular Ca^{++} with anti-CTLA-4. In contradiction to the data presented by Damle et al. [42], Walunas et al. [43] have proposed that CTLA-4 downregulates the proliferative response. Using a monoclonal antibody that appears to block CTLA-4 function they demonstrate an enhancement of T cell proliferation during activation by anti-CD3. Similar results are seen using Fab′ fragments of the antibody, suggesting that the enhanced proliferation is due to interference with the provision of a negative signal. The resolution of these conflicting results awaits further study.

Ligation of CD28 by either a monoclonal antibody or B7-1 expressed on the surface of a CHO cell leads to the induction of a proliferative response in purified human T cells activated with submitogenic doses of anti-CD3 (fig. 4). The provision of the costimulatory signal is effective whether delivered in cis or in trans. The combination of PMA and CHO-B7 also results in proliferation. This response is completely dependent upon B7 as it is abrogated by CTLA4Ig. Similarly, blockade of B7 interactions inhibits the proliferative response in mixed lymphocyte cultures. Thus, B7/CD28 ligand interactions appear to play an important role in regulating the proliferative response of antigen-activated T cells. Given this, it is of interest to examine the consequences of in vivo disruption of the CD28 gene.

CD28-deficient mice have recently been created [44], and initial characterization of

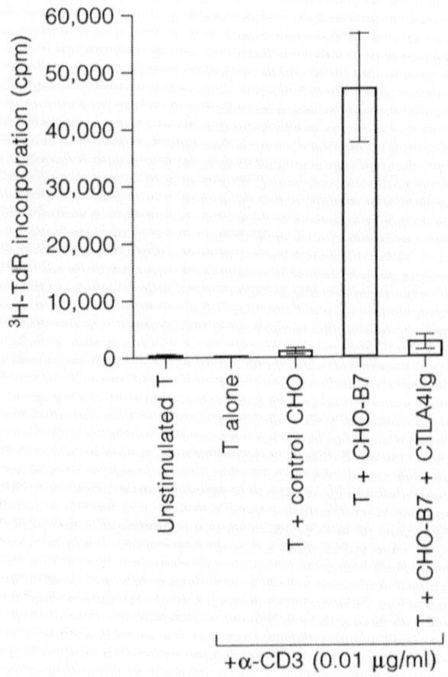

Fig. 4. Stimulation of purified resting human T cells with submitogenic anti-CD3 antibody plus B7-transfected CHO cells leads to proliferation that can be suppressed by CTLA4Ig.

the function of CD28-deficient T cells has provided some interesting insights [9]. Lymph node cells from CD28-deficient mice, activated by an anti-CD3 monoclonal antibody, display a decreased proliferative response in comparison to wild-type cells. Similarly, both alloantigen activation and antigen-specific responses are decreased as compared to wild-type controls. The kinetics of the proliferative response suggests that the initiation of proliferation is intact, but the maintenance of the response is impaired.

A potential explanation for the residual ability of the CD28-deficient cells to proliferate is that costimulation is being provided by CTLA-4, the alternate counterreceptor for B7. In that case, one would expect that B7 blockade would interfere with the proliferative response of CD28-deficient T cells. In fact, inhi-

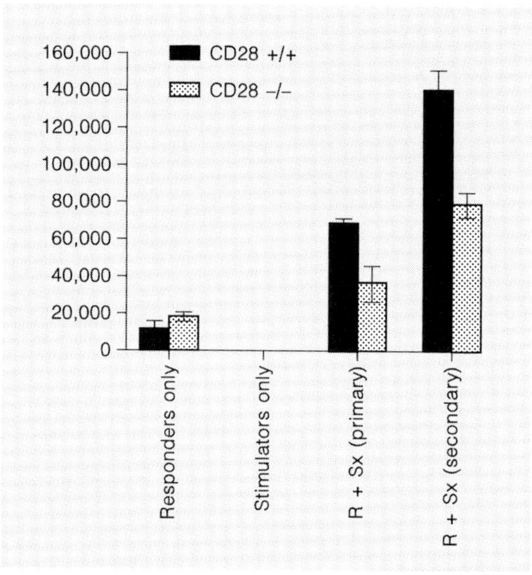

Fig. 5. T cells from wild-type and CD28-deficient mice are capable of both primary and secondary responses to alloantigen challenge. R = Lymph node cells; Sx = irradiated allogeneic cells.

bition of B7/counterreceptor interactions did not alter the proliferation of CD28-deficient T cells in either a positive or negative direction. Thus, it appears that in the absence of CD28, CTLA-4 does not function to either upregulate or decrease the T cells' proliferative response. The data derived from the CD28 knockout mouse is most consistent with CTLA-4 not having any intrinsic signalling capability, but functioning as a high affinity receptor which limits the availability of B7 to interact with CD28.

The two-signal model of T cell activation predicts that in the absence of costimulation, the T cell is rendered anergic. In fact, when T cells from CD28-deficient mice are stimulated with alloantigen, rested and then restimulated, they develop a typical secondary response (fig. 5), albeit somewhat less in magnitude than in wild-type controls. Similarly, when primed with a protein antigen in vivo, lymph node cells from CD28-deficient mice proliferate upon restimulation. Thus, the absence of a CD28-derived second signal does not lead to the induction of anergy, as might have been predicted. This result can be explained by either the presence of non-CD28 costimulatory receptors, or alternatively that the provision of costimulation is not always absolutely required.

When T cells from the CD28-deficient mice were rigorously purified such that no accessory cell activity could be detected, the proliferative response to anti-CD3 stimulation was identical to wild-type controls. In comparison to activation in the presence of accessory cells, the dose-response curve of purified wild-type T cells was shifted to the right, implying that a higher degree of TCR occupancy was required for the initiation of proliferation. Proliferation was not affected by B7 blockade in either wild-type or CD28-deficient T cells. While one cannot formally exclude the possibility that T cells can provide costimulation to each other in trans, these data suggest that TCR signalling alone can be sufficient to induce a proliferative response.

By augmenting the proliferative response of T cells independent of intercellular adhesion and utilizing a signal transduction pathway distinct from the TCR, the provision of a second signal through the B7/CD28 pathway fulfills the requirement for a bona-fide costimulatory event. Whether other receptor ligand interactions will be identified that also meet the necessary criteria remains to be seen.

Conclusions

The T cell surface receptors LFA-1 and CD28 both enhance the proliferative response of an antigen-activated T cell, but through distinct mechanisms. Each can be thought of as a prototype for two important mechanisms of modulating the immune response. The first mechanism involves enhancing the strength of the signal delivered via the TCR. This may be via increasing the physical association between T cell and APC, or by modifying the TCR-derived signal. This mechanism of 'costimulation' is exemplified by LFA-1. Engagement of LFA-1 on the T cell by its ligand on the APC serves to stabilize the interaction between the T cell and accessory cell, which allows for effective engagement of low affinity Ag/MHC complexes on the APC with the TCR. At the present time, there is no conclusive evidence that ligand-binding to LFA-1 provides signals to the T cells that are independent of TCR signalling. Other adhesion molecules may enhance proliferation in a similar way. Schematically, these pathways raise the strength of TCR engagement to the level required for a proliferative response.

In contrast to the adhesion receptors, CD28 appears to function via a novel signal transduction pathway. While the intracellular events remain to be clarified, it seems clear that the signalling is independent of the TCR. Thus while CD28 engagement alone appears to be a neutral event, in concert with antigen-specific activation CD28 ligation results in an increase in the proliferative potential of the T cell, as well as enhanced development of effector function. By costimulating through CD28, a much lower level of TCR engagement is required for successful T cell activation.

At least one other class of costimulatory receptors exists. This class is exemplified by the CD4 and CD8 coreceptors [for recent review, see ref. 45]. These are expressed on the surface of the T cell in close physical association with the TCR. The presence of CD4 confers specificity for class II MHC, and CD8 for class I MHC. The cytoplasmic domain of both CD4 and CD8 associates with the protein tyrosine kinase $p56^{lck}$, which upon activation can phosphorylate the ζ chain of the CD3 complex. Engagement of the TCR causes the CD4 or CD8 coreceptor to colocalize with the TCR. If the binding of CD4 or CD8 to MHC is blocked, much higher levels of antigen are required to signal through the TCR. Thus, the coreceptors probably function both by enhancing the physical interaction of the TCR with the Ag/MHC complex by virtue of their ability to bind to MHC, and restrict interactions to a specific class of MHC molecule. They also modulate signal transduction through interaction with $p56^{lck}$.

Several other putative costimulatory receptors have been identified. While some may act by purely physical mechanisms, others may regulate the signal transduction through the TCR in a more complex manner. CD44, a broadly distributed cell surface receptor that binds hyaluronate, mediates the adhesion of lymphocytes to the high endothelial venules in lymph nodes [46]. In addition, it may also deliver a coactivating signal to the T cell [47]. Similarly, engagement of CD2 can modulate the proliferative response of T cells [for review, see ref. 48], however, signalling through CD2 is dependent upon the presence of the TCR/CD3 complex. Heat-stable antigen has also been suggested as being able to deliver a costimulatory signal to resting murine CD4+ T cells [49]. In part this appears to be dependent on B7/CD28 interactions [50]. Whether HSA can engage a ligand on the T cell and independently deliver a costimulatory remains to be clarified.

While the provision of costimulation is undoubtedly an important mechanism in the regulation of the immune response, it may not be uniformly required. As the studies from the CD28-deficient mice have suggested, under some circumstances TCR engagement alone may be sufficient to induce a T cell proliferative response. Nevertheless, the studies described here serve to underscore the importance of T cell receptors that can regulate the response of T cells encountering an APC. How these interactions serve to modulate in vivo immune responses will undoubtedly be an area of intense research over the next several years.

Acknowledgments

The authors with to thank Therese Conway, Carol McShan, and Patricia Noel for their contribution to this work. J.M.G. is supported in part by grant No. 5F32 AI09042.

References

1 Germain RN, Margulies DH: The biochemistry and cell biology of antigen processing and presentation. Annu Rev Immunol 1993;11:403–450.

2 Bretscher P, Cohn M: A theory of self-nonself discrimination: Paralysis and induction involve the recognition of one and two determinants on an antigen, respectively. Science 1970;169:1042–1049.

3 Lafferty KJ, Cunningham AJ: A new analysis of allogeneic interactions. Aust J Exp Biol Med Sci 1975;53:27–42.

4 Schwartz RH: A cell culture model for T lymphocyte clonal anergy. Science 1990;248:1349–1356.

5 Quill H, Schwartz RH: Stimulation of normal inducer T cell clones with antigen presented by purified Ia molecules in planar lipid membranes: Specific induction of a long-lived state of proliferative nonresponsiveness. J Immunol 1987;138:3704–3712.

6 Jenkins MK, Ashwell JD, Schwartz RH: Allogeneic non-T spleen cells restore the responsiveness of normal T cell clones stimulated with antigen and chemically modified antigen-presenting cells. J Immunol 1988;140:3324–3330.

7 Mueller DL, Jenkins MK, Chiodetti L, Schwartz RH: An intracellular calcium increase and protein kinase C activation fail to initiate T cell proliferation in the absence of a costimulatory signal. J Immunol 1990;144:3701–3709.

8 Linette GP, Hartzman RJ, Ledbetter JA, June CH: HIV-1-infected T cells show a selective signaling defect after perturbation of CD3/antigen receptor. Science 1988;241:573–576.

9 Green JM, Noel PJ, Sperling AI, Walunas TL, Gray GS, Bluestone JA, Thompson CB: Absence of B7-dependent responses in CD28-deficient mice. Immunity 1994;1:508–515.

10 Matsui K, Boniface JJ, Reay PA, Schild H, Fazekas DE, St Groth B, Davis MM: Low affinity interaction of peptide-MHC complexes with T cell receptors. Science 1991;254:1788–1791.

11 Siu G, Hedrick SM, Brian AA: Isolation of the murine intercellular adhesion molecule 1 (ICAM-1) gene: ICAM-1 enhances antigen-specific T cell activation. J Immunol 1989;143:3813–3820.

12 van Noesel C, Miedema F, Brouwer M, de Rie MA, Aarden LA, van Lier RAW: Regulatory properties of LFA-1 α and β chains in human T-lymphocyte activation. Nature 1988;333:850–852.

13 van Seventer GA, Shimizu Y, Horgan KJ, Shaw S: The LFA-1 ligand ICAM-1 provides an important costimulatory signal for T cell receptor-mediated activation of resting T cells. J Immunol 1990;144:4579–4586.

14 Kuhlman P, Moy VT, Lollo BA, Brian AA: The accessory function of murine intercellular adhesion molecule-1 in T lymphocyte activation: Contributions of adhesion and coactivation. J Immunol 1991;146:1773–1782.

15 Damle NK, Klussman K, Aruffo A: Intercellular adhesion molecule-2, a second counter-receptor for CD11a/CD18 (leukocyte function-associated antigen-1), provides a costimulatory signal for T-cell receptor-initiated activation of human T cells. J Immunol 1992;148:665–671.

16 Damle NK, Klussman K, Linsley PS, Aruffo A: Differential costimulatory effects of adhesion molecules B7, ICAM-1, LFA-3, and VCAM-1 on resting and antigen-primed CD4+ T lymphocytes. J Immunol 1992;148:1985–1992.

17 van Seventer GA, Shimizu Y, Horgan KJ, Luce GEG, Webb D, Shaw S: Remote T cell co-stimulation via LFA-1/ICAM-1 and CD2/LFA-3: Demonstration with immobilized ligand/mAb and implication in monocyte-mediated co-stimulation. Eur J Immunol 1991;21:1711–1718.

18 Green JM, Zheng X-G, Shimizu Y, Thompson CB, Turka LA: T cell receptor stimulation, but not CD28 costimulation, is dependent on LFA-1-mediated events. Eur J Immunol 1994;24:265–272.

19 Fischer H, Gjörloff A, Hedlund G, Hedman H, Lundgren E, Kalland T, Sjögren HO, Dohlsten M: Stimulation of human naive and memory T helper cells with bacterial superantigen: Naive CD4+45RA+T cells require a costimulatory signal mediated through the LFA-1/ICAM-1 pathway. J Immunol 1992;148:1993–1998.

20 Marrack P, Kappler J: The staphylococcal enterotoxins and their relatives. Science 1990;248:705–711.

21 Schlievert PM, Shands KN, Dan BB, Schmid GP, Nishimura RD: Identification and characterization of an exotoxin from *Staphylococcus aureus* associated with toxic-shock syndrome. J Infect Dis 1981;143:509–516.

22 Marrack P, Kushnir E, Kappler J: A maternally inherited superantigen encoded by a mammary tumour virus. Nature 1991;349:524–526.

23 Frankel WN, Rudy C, Coffin JM, Huber BT: Linkage of *Mls* genes to endogenous mammary tumour viruses of inbred mice. Nature 1991; 349:526–528.
24 Woodland DL, Happ MP, Gollob KJ, Palmer E: An endogenous retrovirus mediating deletion of αβ T cells? Nature 1991;349:529–530.
25 Dyson PJ, Knight AM, Fairchild S, Simpson E, Tomonari K: Genes encoding ligands for deletion of Vβ11 T cells cosegregate with mammary tumour virus genomes. Nature 1991;349:531–532.
26 Green JM, Turka LA, June CH, Thompson CB: CD28 and staphylococcal enterotoxins synergize to induce MHC-independent T-cell proliferation. Cell Immunol 1992;145: 11–20.
27 van Seventer GA, Bonvini E, Yamada H, Conti A, Stringfellow S, June CH, Shaw S: Costimulation of T cell receptor/CD3-mediated activation of resting human $CD4^+$ T cells by leukocyte function-associated antigen-1 ligand intercellular cell adhesion molecule-1 involves prolonged inositol phospholipid hydrolysis and sustained increase of intracellular Ca^{2+} levels. J Immunol 1992;149:3872–3880.
28 Gmünder H, Lesslauer W: A 45-kDa human T-cell membrane glycoprotein functions in the regulation of cell proliferative responses. Eur J Biochem 1984;142:153–160.
29 June CH, Bluestone JA, Nadler LM, Thompson CB: The B7 and CD28 receptor families. Immunol Today 1994;15:321–331.
30 June CH, Ledbetter JA, Gillespie MM, Lindsten T, Thompson CB: T-cell proliferation involving the CD28 pathway is associated with cyclosporine-resistant interleukin 2 gene expression. Mol Cell Biol 1987; 7:4472–4481.
31 Vandenberghe P, Freeman GJ, Nadler LM, Fletcher MC, Kamoun M, Turka LA, Ledbetter JA, Thompson CB, June CH: Antibody and B7/BB1-mediated ligation of the CD28 receptor induces tyrosine phosphorylation in human T cells. J Exp Med 1992;175:951–960.
32 Linsley PS, Clark EA, Ledbetter JA: T-cell antigen CD28 mediates adhesion with B cells by interacting with activation antigen B7/BB-1. Proc Natl Acad Sci USA 1990;87:5031–5035.
33 Gimmi CD, Freeman GJ, Gribben JG, Sugita K, Freedman AS, Morimoto C, Nadler LM: B-cell surface antigen B7 provides a costimulatory signal that induces T cells to proliferate and secrete interleukin 2. Proc Natl Acad Sci USA 1991;88:6575–6579.
34 Freeman GJ, Gribben JG, Boussiotis VA, Ng JW, Restivo VA Jr, Lombard LA, Gray GS, Nadler LM: Cloning of B7-2: A CTLA-4 counter-receptor that costimulates human T cell proliferation. Science 1993;262: 909–911.
35 Azuma M, Ito D, Yagita H, Okumura K, Phillips JH, Lanier LL, Somoza C: B70 antigen is a second ligand for CTLA-4 and CD28. Nature 1993;366:76–79.
36 Nickoloff BJ, Mitra RS, Lee K, Turka LA, Green J, Thompson C, Shimizu Y: Discordant expression of CD28 ligands, BB-1, and B7 on keratinocytes in vitro and psoriatic cells in vivo. Am J Pathol 1993;142: 1029–1040.
37 Symington FW, Brady W, Linsley PS: Expression and function of B7 on human epidermal Langerhans cells. J Immunol 1993;150:1286–1295.
38 Linsley PS, Brady W, Urnes M, Grosmaire LS, Damle NK, Ledbetter JA: CTLA-4 is a second receptor for the B cell activation antigen B7. J Exp Med 1991;174:561–569.
39 Lenschow DJ, Zeng Y, Thistlethwaite JR, Montag A, Brady W, Gibson MG, Linsley PS, Bluestone JA: Long-term survival of xenogeneic pancreatic islet grafts induced by CTLA4Ig. Science 1992;257:789–792.
40 Turka LA, Linsley PS, Lin H, Brady W, Leiden JM, Wei R-Q, Gibson ML, Zheng X-G, Myrdal S, Gordon D, Bailey T, Bolling SF, Thompson CB: T-cell activation by the CD28 ligand B7 is required for cardiac allograft rejection in vivo. Proc Natl Acad Sci USA 1992;89:11102–11105.
41 Linsley PS, Wallace PM, Johnston J, Gibson MG, Greene JL, Ledbetter JA, Singh C, Tepper MA: Immunosuppression in vivo by a soluble form of the CTLA-4 T cell activation molecule. Science 1992;257: 792–795.
42 Damle NK, Klussman K, Leytze G, Myrdal S, Aruffo A, Ledbetter JA, Linsley PS: Costimulation of T lymphocytes with integrin ligands intercellular adhesion molecule-1 or vascular cell adhesion molecule-1 induces functional expression of CTLA-4, a second receptor for B7. J Immunol 1994;152:2686–2697.
43 Walunas TL, Lenschow DJ, Bakker CY, Linsley PS, Freeman GJ, Green JM, Thompson CB, Bluestone JA: CTLA-4 can function as a negative regulator of T cell activation. Immunity 1994;1:405–413.
44 Shahinian A, Pfeffer K, Lee KP, Kündig TM, Kishihara K, Wakeham A, Kawai K, Ohashi PS, Thompson CB, Mak TW: Differential T cell costimulatory requirements in CD28-deficient mice. Science 1993;261:609–612.
45 Janeway CA Jr: The T cell receptor as a multicomponent signalling machine: CD4/CD8 coreceptors and CD45 in T cell activation. Annu Rev Immunol 1992;10:645–674.
46 Aruffo A, Stamenkovic I, Melnick M, Underhill CB, Seed B: CD44 is the principal cell surface receptor for hyaluronate. Cell 1990;61:1303–1313.
47 Huet S, Groux H, Caillou B, Valentin H, Prieur A-M, Bernard A: CD44 contributes to T cell activation. J Immunol 1989;143:798–801.
48 Bierer BE, Burakoff SJ: T-lymphocyte activation: The biology and function of CD2 and CD4. Immunol Rev 1989;111:267–294.
49 Liu Y, Jones B, Aruffo A, Sullivan KM, Linsley PS, Janeway CA Jr: Heat-stable antigen is a costimulatory molecule for CD4 T cell growth. J Exp Med 1992;175:437–445.
50 Liu Y, Jones B, Brady W, Janeway CA Jr, Linley PS: Co-stimulation of murine CD4 T cell growth: Cooperation between B7 and heat-stable antigen. Eur J Immunol 1992;22: 2855–2859.

Immunol Res 1994;13:244–252

Jim Miller

Department of Molecular Genetics and Cell Biology, University of Chicago, Ill., USA

Endosomal Localization of MHC Class II-Invariant Chain Complexes

Key Words
Major histocompatibility complex class II invariant chain
Antigen processing
Protein targeting
Intracellular sorting signals

Abstract
CD4-positive T cells recognize foreign antigens displayed on the surface of antigen-presenting cells as small peptides bound to MHC class II molecules. Thus, the ability of antigen-presenting cells to generate these class II-peptide complexes is central to the initiation and regulation of immune responses. Class II predominantly associates with peptides derived from soluble protein antigens that are internalized and degraded within endosomal compartments. It is within these endosomal compartments that class II encounters and binds antigenic peptides. A number of signals have been implicated in directing the intracellular transport of class II to endosomes. These include sequences within class II itself and within the class II-associated invariant chain (Ii).

Intracellular Protein Targeting

Protein transport through the secretory system is for the most part a unidirectional process. Plasma membrane and secreted proteins are cotranslationally translocated into the endoplasmic reticulum (ER), where they undergo folding and oligomerization steps. Protein folding in the ER is a complex process that requires facilitation by a series of protein cofactors, called chaperones. Chaperones associate with partially folded proteins in the ER, effectively stabilizing intermediates that otherwise would misfold and aggregate. Once proteins are folded they exit the ER and transverse the Golgi apparatus. Within the Golgi stacks, modifications of N-linked carbohydrate take place in a sequential series of reactions that are thought to be associated with distinct cisternae of the Golgi (cis, medial, and trans). The final compartment in the Golgi is the trans-Golgi network (TGN), where plasma membrane and secreted proteins exit to the cell surface.

Jim Miller
Department of Molecular Genetics and Cell Biology, University of Chicago
920 East 58th Street
Chicago, IL 60637 (USA)

0257–277X/94/
0134–0244$8.00/0

The current view is that this directionality is determined by a process of bulk flow, where proteins can simply follow a default membrane transport pathway from the ER through the Golgi and on to the plasma membrane. Layered on top of this constitutive pathway are two mechanisms that account for localization of proteins to specific compartments, retention signals and targeting signals. (1) Retention signals function by blocking further transport along the default pathway, effectively localizing a protein within a specific compartment. Two mechanisms of retention have been described. First, ER retention of chaperones and their associated partially folded proteins is maintained by retrieval from the cis-Golgi [1]. These proteins have a C-terminal, lumenal sequence, KDEL, that is recognized by a specific cis-Golgi receptor. The receptor-ligand complex then recycles back to the ER, where the chaperone is released. A second method of retention has been described for resident transmembrane proteins in the Golgi. In this case, the retention signal is encoded by the transmembrane segment and retention is either imparted by transmembrane aggregation or changes in the physical dimensions of the membrane [2].

(2) Targeting signals direct the transport of proteins out of the constitutive pathway into a different compartment or membrane system. Protein transport between most intracellular membrane compartments is mediated by vesicular transport. Small membrane vesicles bud from the donor compartment, specifically bind to the surface of an acceptor compartment, and membrane fusion is induced to allow transport of soluble protein cargo and associated transmembrane proteins. What controls the specificity of these vesicular targeting and fusion events is not fully understood, although several key components of this process have been identified [3]. Protein targeting appears to occur primarily at specific branch points along the constitutive pathway. As mentioned above the cis-Golgi compartment is involved in sorting proteins from the constitutive pathway to be recycled back to the ER. The other major sorting compartment in the secretory pathway is the TGN. Here, proteins can follow the default pathway to the cell surface, can be packaged and stored into secretory vesicles that are released in a regulated fashion (for example, hormones), can be targeted directly to late endosomes and lysosomes, or in polarized cells can be transported to either the apical or basolateral surfaces. Specific protein targeting is mediated by segregating proteins into different transport vesicles. Although the majority of targeting signals that have been described are encoded within the cytosolic tail of the target protein itself, transmembrane and lumenal domain signal have also been implicated, that may function independently or by association with a specific receptor.

Protein transport through the endocytic pathway follows the same basic principles as the secretory pathway. However, endocytic compartments do not exist as stable, morphologically discrete membrane compartments, as is the Golgi. This creates some confusion with defining the specific transport routes through this pathway. In general, proteins are internalized from the cell surface into the early endosome. The early endosome is the primary sorting compartment within the endosomal pathway. Here, the constitutive route for membrane proteins appears to be recycling back to the cell surface [4]. Soluble ligands are fractionated away from membrane proteins through a series of vesicular transport events and are passively transported to late endosomes. Transport of membrane proteins from the early to late endosomes typically requires the presence of specific targeting signals, although recycling proteins can be diverted into this pathway by cross-linking [5, 6]. Interesting, it has recently been shown that some proteins can be specifically targeted from the early endosome to the TGN [7, 8], indicating that the exocytic and endocytic pathways are more interconnected than once thought. Proteins that arrive in late endosomes typically follow a constitutive route to lysosomes, although there is clear evidence for targeting of ligand-free mannose phosphate receptor from the late endosome back to the TGN.

MIIC Compartment

Unlike most plasma membrane proteins that follow the constitutive route to the cell surface, class II is targeted and retained within endosomal compartments prior to stable expression at the cell surface [9]. Although class II can be found throughout the endosomal pathway [10–13], it predominantly resides within an unusual vesicle, called the MIIC [14]. This compartment was initially characterized by electron microscopy as containing high levels of class II, lots of internal membranes, often referred to as multivesicular bodies, and the endosomal/lysosomal markers lamp-1 and CD63. The MIIC is unique from typical endosomes in that it contained little or no transferrin receptor, a marker for early endosomes, or mannose phosphate receptor, a marker for late endosomes, and did not accumulate fluid phase endocytosed material, an indicator of lysosomes. Recently, the intracellular class II compartment has been characterized following subcellular fractionation [15–18]. Although there are subtle differences in the various compartments identified, these most likely represent differences between the specific cells tested and the relative sensitivity of the assay systems. Thus, the growing consensus is that intracellular class II is predominantly, although not exclusively, localized in a subcompartment of late endosomes/lysosomes, called the MIIC.

The MIIC appears to be a functionally relevant compartment for class II biosynthesis. The MIIC, or a similar compartment, is increased in size by overexpression of class II and/or Ii [12, 19, 20]. It is also increased in size [21] in cell lines that lack HLA-DM expression and are defective in the generation of class II-peptide complexes [22, 23]. Finally, it is the site where SDS-stable class II dimers are first detected during class II biosynthesis [15–18]. Stabilization of class II to SDS is an indicator of class II-peptide binding. Thus, it appears that MIIC may be the compartment where class II is targeted to and retained within until it can effectively load with antigenic peptide.

Route from the TGN to the MIIC

Although progress has been made in identifying the relevant endosomal compartments, the route the class II-Ii complexes travel to arrive in this compartment remains unresolved. The original interpretation of the data suggested that class II-Ii complexes were targeted directly from the TGN to endosomes. This was based on kinetic data that the complex could be detected in endosomes prior to stable integration into the plasma membrane. However, kinetic data cannot rule out a transient cell surface intermediate that gains access to endosomes following rapid internalization.

Support for the direct endosomal transport model comes from electron microscopic studies that demonstrate that in the TGN, class II is segregated into separate transport vesicles from proteins following the constitutive pathway to the plasma membrane [14]. Interestingly, class II is also segregated from mannose phosphate receptor which is transported directly from the TGN to late endosomes. These structural data suggest that class II is targeted from the TGN into an endosomal compartment that is distinct from conventional late endosomes. This raises the possibility that class II goes directly from the TGN to the MIIC compartment. Support for the plasma membrane intermediate model comes from biochemical studies on the internalization of class II-Ii complexes from the cell surface [24]. In human B cell lines that express high levels of class II and Ii, a small number of class II-Ii complexes can be detected at the cell surface. These complexes are rapidly internalized at a rate that suggests they may be a natural intermediate in the biosynthesis of class II. The difficulty with these analyses is that in order to detect the transient intermediate, high levels of expression are required. This raises the possibility that the cell surface pool of class II-Ii complexes is not a natural intermediate, but rather represents an overflow of molecules that have saturated the normal sorting machinery in the TGN and so follow the bulk flow to the surface. This problem has been encountered in several other systems and at present is not fully resolvable [25].

Although in a sense the exact route of transport is somewhat academic, it is relevant to two issues. The first is the specific compartments where class II might access peptide. Although the MIIC may be the predominant site for class II-peptide binding, there may be distinct advantages to class II gaining access to peptide in a variety of endosomal compartments because different peptide epitopes might be released under the different proteolytic/denaturing conditions of different compartments. This would allow presentation of a broader array of peptides to T cells. Second, from a cell biology perspective, understanding the route of intracellular transport will provide clues into where and how various intracellular targeting signals might function.

Endosomal Localization Signals

Much of the focus on endosomal localization of the class II-Ii complex has centered on Ii. Interestingly several different signals have been identified in Ii as well as in class II itself. Thus, targeting appears to be controlled by several different signals that may be functionally redundant or may function independently at different stages in transport.

Invariant Chain

Several studies have indicated that Ii can enhance the localization of class II into the endocytic compartment [12, 26–31]. Expression of class II in the absence of Ii results in a predominant ER staining pattern, with only a small vesicular component. Coexpression with Ii drives class II out of the ER and increases the relative amount of class II in vesicular structures. This dramatic change in steady-state localization of class II in the presence and absence of Ii can be attributed to three functions of Ii. First, Ii can promote class II assembly and folding in the ER, effectively increasing the efficiency of transport from the ER to the Golgi [29–34]. Second, Ii contains a dominant endosomal targeting signal directing the transport of class II into endosomal compartments (see below). Finally, Ii can retain class II in endosomes, increasing the dwell time, and dramatically affecting the steady-state localization of the class II-Ii complex [35, 36].

The Ii targeting/retention functions have been mapped primarily to the cytosolic tail and transmembrane regions of Ii, but some lumenal sequences may also be involved (see below). Initial truncation studies indicated that the first 15 amino acids of the Ii cytosolic tail encoded an endosomal localization signal [37, 38]. Although originally analyzed in the absence of class II, expression of truncation mutants of Ii in the presence of class II results in the cell surface expression of class II-Ii complexes [39, 40]. When the 15-amino acid sequence was analyzed in greater detail, two independent di-leucine-like motifs were identified [41, 42]. Di-leucine motifs were first identified in regulating the lysosomal targeting of CD3 γ-chain [43] and have been shown to regulate both TGN-lysosomal targeting and/or plasma membrane internalization of several transmembrane proteins. The leucine-isoleucine at amino acids 7–8 is clearly critical for Ii transport. The second di-leucine-like motif is equally important, but less well defined and appears to be located within amino acids 15–17. Interestingly, we have found that this signal requires multimerization to function [44]. Intracellularly, the class II-Ii complex is actually a nine-chain complex, consisting of an Ii trimer and three class II heterodimers [45, 46]. Complexes that contain two or three intact Ii cytosolic domains are targeted efficiently to endosomes and are not detected at the cell surface. In contrast, complexes that contain only a single intact Ii cytosolic domain are readily found in the plasma membrane [44]. Thus, whether the Ii cytosolic tail functions as an internalization signal from the cell surface or as a TGN to endosome targeting signal, a single signal is not sufficient.

Although the two di-leucine-like motifs are necessary for Ii targeting, two recent studies using chimeras between Ii and transferrin receptor (TfR) suggest that they are not sufficient. First, when the Ii cytosolic tail and transmembrane regions were transplanted onto the TfR, the chimeric TfR-Ii molecules were efficiently targeted to endosomes and rapidly degraded [41]. However, when the Ii cytosolic tail alone was transferred, the efficiency of sorting and the degradation rate was reduced, suggesting that the transmembrane region (or the context that the Ii cytosolic tail was expressed) could regulate Ii transport. Similar conclusions were reached when the reciprocal chimeric protein, containing the lumenal and transmembrane sequences of Ii attached to the TfR cytosolic tail, was analyzed [K. Swier and J. Miller, manuscript in preparation]. When expressed with class II, the Ii-TfR chimera was rapidly transported to the cell surface, rapidly internalized, and recycled. The Ii-TfR chimera was eventually degraded, but with a half-life of about 8 h, compared to 1–2 h for wild-type Ii and 12–16 h for class II. In contrast, class II that was associated with truncated Ii, that lacked amino acids 2–17 containing the two di-leucine-like motifs, was also transported rapidly to the cell surface, but then slowly internalized and rapidly degraded (with a half-life of 3–4 h) [39]. These results implicated a role for additional signals in the Ii cytosolic tail within residues 18–29. Taken together, these findings indicate that multiple signals in the Ii cytosolic tail and transmembrane regions interact to allow for efficient localization of class II-Ii complexes to endosomes.

Class II

Although endosomal localization of class II can be enhanced by association with Ii, class II itself can also gain access to endosomes. This was initially suggested by the ability of Ii-negative cells to process and present some exogenous antigens to T cells [31, 47–49]. This functional readout implied that class II could gain access to the relevant peptide loading compartments, for at least a subset of antigenic peptides. This possibility was confirmed by a number of morphological studies that clearly identified class II within late endosomal compartments [27, 28, 50]. Recently, the ability of class II to localize to endosomes has been mapped to a lumenal sequence in the class II beta chain encoded by amino acids 80–84 [27]. Highly conservative mutations within this region dramatically modulate the relative localization of class II to endosomal compartments. Thus, in addition to the sequences in the Ii cytosolic tail and transmembrane regions, lumenal signals in class II itself can determine the efficiency of targeting the class II-Ii complex into endosomes.

Sorting within Endosomes

The dramatic effects of Ii on class II folding, peptide loading and endosomal localization would suggest that Ii should enhance antigen presentation. Although Ii can increase the efficiency of antigen presentation as much as 100- to 1,000-fold, it does so only for a subset of antigenic epitopes [29, 31, 47, 49, 51]. It is not clear whether this reflects the relative access of class II to different intracellular compartments where various antigenic peptides might be generated, the relative ability of antigenic peptide to associate with class II, or the relative requirement for protein cofactors, such as HLA-DM. Surprisingly, in those cases where Ii can enhance antigen presentation, this effect is largely imparted by the alternatively spliced p41 form of Ii [49]. The two isoforms of Ii, p31 and p41, are identical through the cytosolic tail and transmembrane regions involved in sorting and the CLIP region that has been implicated in class II association, folding, and inhibition of peptide binding [52, 53]. The two molecules differ by the inclusion of a 64-amino acid segment near the carboxy terminus of p41. This region appears to play a specialized role in some aspect of antigen presentation that is distinct from the shared functions between p31 and p41.

Recently, we have found a difference in the post-Golgi processing of class II-associated p31 and p41 [54]. Both forms of Ii are targeted to endosomes and degraded with the same kinetics. However, p31 degradation results in a rapid dissociation from class II, while p41 is degraded into an amino-terminal 12-kD fragment that remains associated with class II for several hours. This altered degradation of p31 and p41 could result either from a change in conformation that alters sensitivity to proteases or to an alteration in endosomal localization that allows access of the class II-Ii complexes to a different set of proteases. Although evidence for the first model has recently been presented [20], we tend to favor the second model. This is based on the finding that p41 can alter the degradation pattern of p31 associated in the same class II-Ii complex [54]. Although this could still reflect a conformational change in protease sensitivity, this change would have to be transduced throughout the Ii trimer. Alternatively, p41 may contain a dominant endosomal targeting signal that regulates access of class II-Ii complexes to a specific endosome, with an altered concentration of proteases, allowing for altered degradation of all associated Ii molecules, p31 and p41 alike. Although additional studies are required to distinguish these possibilities, these studies raise the possibility that different signals might regulate entry of the class II-Ii complex into the endocytic pathway and localization to specific compartments within this pathway, such as the MIIC.

Summary and Perspectives

Over the past few years, we have learned a great deal about protein targeting in general and about the specific endosomal localization signals in the class II-II complex itself. One feature or class II transport, and of transport of oligomeric protein complexes in general, is the apparently large number of individual sorting signals that are present in the complex. One rationale for this observation is that protein folding and oligomeric assembly is a relatively low fidelity process, in comparison to nucleic acid biochemistry in particular. Thus, a series of quality control mechanisms have developed to regulate protein biosynthesis and transport, such that misassembled or partially assembled complexes do not arrive at the cell surface and compete with complete functionally assembled complexes. This is best understood for the TcR-CD3 complex, where the level of expression of the entire complex can be regulated by the concentration of a single chain [55]. This results in the production of various partial complexes, that are sorted and/or retained in different compartments prior to arrival at the cell surface. Each chain of the TcR-CD3 complex contains its own independent signals that when present in complexes with other chains can be masked or highlighted. A similar level of control may function for class II-Ii complexes, where depending on the constituency of the complex, class II alone, Ii alone, and class II-Ii complexes that do and do not contain p41 could have different intracellular fates. Alternatively, because the class II-Ii complex transverses much of the exocytic and endocytic pathways in cells, different signals might function at different stages in class II biosynthesis. Clearly, there is much more to understand.

References

1 Pelham H: Recycling of proteins between the ER and Golgi complex. Curr Opin Cell Biol 1991;3:585–591.

2 Bretscher M, Munro S: Cholesterol and the Golgi apparatus. Science 1993;261:1280–1281.

3 Ferro-Novick S, Novick P: The role of GTP-binding proteins in transport along the exocytic pathway. Annu Rev Cell Biol 1993;9:575–599.

4 Mayor S, Presley J, Maxfield F: Sorting of membrane components from endosomes and subsequent recycling to the cell surface occurs by a bulk flow process. J Cell Biol 1993;121:1257–1269.

5 Davis C, Goldstein J, Sudhof T, Anderson R, Russell D, Brown M: Acid-dependent ligand dissociation and recycling of LDL receptor mediated by growth factor homology region. Nature 1987;326:760–765.

6 Tabas I, Myers J, Innerarity T, Xu X, Arnold K, Boyles J, Maxfield F: The influence of particle size and multiple apoprotein-receptor interactions on the endocytic targeting of β-VLDL in mouse peritoneal macrophages. J Cell Biol 1991;115:1547–1560.

7 Molloy S, Thomas L, VanSlyke J, Stenberg P, Thomas G: Intracellular trafficking and activation of the furin proprotein convertase: Localization to the TGN and recycling from the cell surface. EMBO J 1994;13:18–33.

8 Humphrey J, Peters P, Yuan L, Bonifacino J: Localization of TGN38 to the trans-Golgi network: Involvement of a cytoplasmic tyrosine-containing sequence. J Cell Biol 1993;120:1123–1135.

9 Neefjes J, Stollorz V, Peters P, Geuze H, Ploegh H: The biosynthetic pathway of MHC class II but not class I molecules intersects the endocytic route. Cell 1990;61:171–183.

10 Guagliardi L, Koppelmann B, Blum J, Marks M, Cresswell P, Brodsky F: Colocalization of molecules involved in antigen processing and presentation in an early endocytic compartment. Nature 1990;343:133–139.

11 Pieters J, Horstmann H, Bakke O, Griffiths G, Lipp J: Intracellular transport and localization of major histocompatibility complex class II molecules and associated invariant chain. J Cell Biol 1991;115:1213–1223.

12 Romagnoli P, Layet C, Yewdell J, Bakke O, Germain R: Relationship between invariant chain expression and major histocompatibility complex class II transport into early and late endocytic compartments. J Exp Med 1993;177:583–596.

13 Harding C, Geuze H: Class II MHC molecules are present in macrophage lysosomes and phagolysosomes that function in the phagocytic processing of *Listeria monocytogenes* for presentation to T cells. J Cell Biol 1992;119:531–542.

14 Peters P, Neefjes J, Oorschot V, Ploegh H, Geuze H: Segregation of MHC class II molecules from MHC class I molecules in the Golgi complex for transport to lysosomal compartments. Nature 1991;349:669–676.

15 Amigorena S, Drake J, Webster P, Mellman I: Transient accumulation of new class II MHC molecules in a novel endocytic compartment in B lymphocytes. Nature 1994;369:113–120.

16 Qiu Y, Xu X, Wandinger-Ness W, Calke D, Pierce S: Separation of subcellular compartments containing distinct functional forms of MHC class II. J Cell Biol 1994;125:595–605.

17 Tulp A, Verwoerd D, Dobberstein B, Ploegh H, Pieters J: Isolation and characterization of the intracellular MHC class II compartment. Nature 1994;369:120–126.
18 West M, Lucocq J, Watts C: Antigen processing and class II MHC peptide-loading compartments in human B-lymphoblastoid lines. Nature 1994;369:147–151.
19 Calafat J, Nijenhuis M, Janssen H, Tulp A, Dusseljee S, Wubbolts R, Neefjes J: Major histocompatibility complex class II molecules induce the formation of endocytic MIIC-like structures. J Cell Biol 1994;126: 967–977.
20 Arunachalam B, Lamb C, Cresswell P: Transport properties of free and MHC class II-associated oligomers containing different isoforms of human invariant chain. Int Immunol 1994;6:439–451.
21 Riberdy J, Avva R, Geuze H, Cresswell P: Transport and intracellular distribution of MHC class II molecules and associated invariant chain in normal and antigen-processing mutant cell lines. J Cell Biol 1994; 126:1225–1237.
22 Fling S, Arp B, Pious D: HLA-DMA and HLA-DMB genes are both required for MHC class II/peptide complex formation in antigen presenting cells. Nature 1994;368:554–558.
23 Morris P, Shaman J, Attaya M, Amaya M, Goodman S, Bergman C, Monaco J, Mellins E: An essential role for HLA-DM in antigen presentation by class II MHC molecules. Nature 1994;368:551–554.
24 Roche P, Teletski C, Stang E, Bakke O, Long E: Cell surface HLA-DR-invariant chain complexes are targeted to endosomes by rapid internalization. Proc Natl Acad Sci USA 1993;90:8581–8585.
25 Harter C, Mellman I: Transport of the lysosomal membrane glycoprotein lgp120 (lgp-A) to lysosomes does not require appearance on the plasma membrane. J Cell Biol 1992; 117:311–325.
26 Lamb C, Yewdell J, Bennink J, Cresswell P: Invariant chain targets HLA class II molecules to acidic endosomes containing internalized influenza virus. Proc Natl Acad Sci USA 1991;88:5998–6002.
27 Chervonsky A, Gordon L, Sant A: A segment of the MHC class II β chain plays a critical role in targeting class II molecules to the endocytic pathway. Int Immunol 1994;6:973–982.
28 Simonsen A, Momburg F, Drexler J, Hammerling G, Bakke O: Intracellular distribution of the MHC class II molecules and the associated invariant chain in different cell lines. Int Immunol 1993;5:903–917.
29 Bikoff E, Huang L, Episkopou V, Meerwijk J, Germain R, Robertson E: Defective major histocompatibility complex class II assembly, transport, peptide acquisition, and CD4+ T cell selection in mice lacking invariant chain. J Exp Med 1993;177: 1699–1712.
30 Elliott E, Drake J, Amigorena S, Elsemore J, Webster P, Mellman I, Flavell R: The invariant chain is required for intracellular transport and function of major histocompatibility complex class II molecules. J Exp Med 1994;179:681–694.
31 Viville S, Neefjes J, Lotteau V, Dierich A, Lemeur M, Ploegh H, Benoist C, Mathis D: Mice lacking the MHC class II-associated invariant chain. Cell 1993;72:635–648.
32 Anderson M, Miller J: Invariant chain can function as a chaperone protein for class II major histocompatibility complex molecules. Proc Natl Acad Sci USA 1992;89:2282–2286.
33 Layet C, Germain R: Invariant chain promotes egress of poorly expressed, haplotype-mismatched class II major histocompatibility complex AαAβ dimers from the endoplasmic reticulum/cis-Golgi compartment. Proc Natl Acad Sci USA 1991;88:2346–2350.
34 Schaiff W, Hruska K, Bono C, Shuman S, Schwartz B: Invariant chain influences post-translational processing of HLA-DR molecules. J Immunol 1991;147:603–608.
35 Loss G, Sant A: Invariant chain retains MHC class II molecules in the endocytic pathway. J Immunol 1993;150:3187–3197.
36 Neefjes JJ, Ploegh HL: Inhibition of endosomal proteolytic activity by leupeptin blocks surface expression of MHC class II molecules and their conversion to SDS resistant αβ heterodimers in endosomes. EMBO J 1992;11:411–416.
37 Bakke O, Dobberstein B: MHC class II-associated invariant chain contains a sorting signal for endosomal compartments. Cell 1990;63:707–716.
38 Lotteau V, Teyton L, Peleraux A, Nilsson T, Karlsson L, Schmid S, Quaranta V, Peterson P: Intracellular transport of class II MHC molecules directed by invariant chain. Nature 1990;348:600–605.
39 Anderson M, Swier K, Arneson L, Miller J: Enhanced antigen presentation in the absence of the invariant chain endosomal localization signal. J Exp Med 1993;178:1959–1969.
40 Roche P, Teletski C, Karp D, Pinet V, Bakke O, Long E: Stable surface expression of invariant chain prevents peptide presentation by HLA-DR. EMBO J 1992;11:2841–2847.
41 Odorizzi C, Trowbridge I, Xue L, Hopkins C, Davis C, Collawn J: Sorting signals in the MHC class II invariant chain cytoplasmic tail and transmembrane region determine trafficking to an endocytic processing compartment. J Cell Biol 1994; 126:317–330.
42 Pieters J, Bakke O, Dobberstein B: The MHC class II-associated invariant chain contains two endosomal targeting signals within its cytoplasmic tail. J Cell Sci 1993;106:831–846.
43 Letourneur F, Klausner R: A novel di-leucine motif and a tyrosine-based motif independently mediate lysosomal targeting and endocytosis of CD3 chains. Cell 1992;69:1143–1157.
44 Arneson L, Miller J: Invariant chain multimerization is necessary for efficient endosomal localization of class-invariant chain complexes. Submitted 1994.
45 Roche P, Marks M, Cresswell P: Formation of a nine subunit complex by HLA class II glycoproteins and the invariant chain. Nature 1991;354:392–394.
46 Lamb C, Cresswell P: Assembly and transport properties of invariant chain trimers and HLA-DR-invariant chain complexes. J Immunol 1992;148:3478–3482.

47 Nadimi F, Moreno J, Momburg F, Heuser A, Fuchs S, Adorini L, Hammerling G: Antigen presentation of hen egg lysozyme but not ribonuclease A is augmented by the MHC class II associated invariant chain. Eur J Immunol 1991;21:1255–1263.
48 Peterson M, Miller J: Invariant chain influences the immunological recognition of MHC class II molecules. Nature 1990;345:172–174.
49 Peterson M, Miller J: Antigen presentation enhanced by the alternatively spliced invariant chain gene product p41. Nature 1992;357:596–598.
50 Salamero J, Humbert M, Cosson P, Davoust J: Mouse B lymphocyte specific endocytosis and recycling of MHC class II molecules. EMBO J 1990;9:3489–3496.
51 Stockinger B, Pessara U, Lin R, Habicht J, Grez M, Koch N: A role of Ia-associated invariant chains in antigen processing and presentation. Cell 1989;56:683–689.
52 Romagnoli P, Germain R: The CLIP region of invariant chain plays a critical role in regulating major histocompatibility complex class II folding, transport, and peptide occupancy. J Exp Med 1994;180:1107–1113.
53 Freisewinkel I, Schneck K, Koch N: The segment of invariant chain that is critical for association with major histocompatibility complex class II molecules contains the sequence of peptide eluted from class II polypeptides. Proc Natl Acad Sci USA 1993;90:9703–9706.
54 Fineschi B, Arneson L, Naujokas M, Miller J: Differential proteolysis of invariant chain trimers containing a single p41 invariant chain component. Submitted 1994.
55 Klausner R, Lippincott-Schwartz J, Bonifacino J: The T cell antigen receptor: Insights into organelle biology. Annu Rev Cell Biol 1990;6:403–431.

Immunol Res 1994;13:253–267

Andrea J. Sant

Department of Pathology,
Committee on Immunology,
University of Chicago, Ill., USA

Endogenous Antigen Presentation by MHC Class II Molecules

Key Words
Major histocompatibility complex class II
Antigen processing
Intracellular trafficking
Antigen-presenting cell
Invariant chain

Abstract
T cell recognition of antigen requires that a complex form between peptides derived from the protein antigen and cell surface glycoproteins encoded by genes within the major histocompatibility complex (MHC). MHC class II molecules present both extracellular (exogenous) and internally synthesized (endogenous) antigens to the CD4 T cell subset of lymphocytes. The mechanisms of endogenous antigen presentation are the subject of this review. Isolation and amino acid sequencing of peptides bound to the class II molecule indicate that a very high proportion (70–90%) of the total peptides presented by the class II molecule are in fact derived from the pool of proteins that are synthetized within the antigen-presenting cell (APC). This type of sequence information as well as the study of model antigens has indicated that proteins expressed in a diversity of intracellular sites, including the cell surface, endoplasmic reticulum and cytosol can gain access to the class II molecule, albeit with different efficiencies. The main questions that remain to be answered are the intracellular trafficking patterns that allow colocalization of internally synthesized antigens with the class II molecule, the site(s) within the cell where peptide:class II molecule complex formation can take place and whether presentation of 'foreign' as well as 'self' antigens takes place by mechanisms that vary from one cell type to another or that vary with the metabolic state of the APC. If such variability exists, is would imply that the array of peptides displayed by class II molecules at the cell surface has similar variability, a possibility that would impact on self tolerance and autoimmunity.

Dr. Andrea J. Sant
Department of Pathology, Committee on Immunology
University of Chicago
5841 S. Maryland Avenue MC1089
Chicago, IL 60637 (USA)

0257–277X/94/
0134–0253$8.00/0

Introduction

T cells play a critical regulatory role in the immune system. Accordingly, there has been great interest in understanding the mechanisms that are involved in T cell recognition. Work by a number of investigators has established that the fundamental event that allows T cell recognition of antigen to occur is the formation of a complex between peptides derived from the antigen and intact cell surface proteins encoded for by genes within the major histocompatibility complex (MHC). Two classes of MHC molecules, MHC class I and class II, are specialized for presentation of antigenic peptides for recognition by T cells. Crystallographic evidence reveals a peptide binding groove at the distal terminus of the MHC class I molecules [1, 2]. A similar binding pocket has been observed in MHC class II crystals [3, 4] that is constructed from the amino-terminal, genetically polymorphic domains of the class II α and β chains.

Although their structural domains are similar, the cellular compartments in which MHC class I and class II molecules function appear to differ. MHC class II molecules have the specialized capacity to present peptides generated from extracellular antigens which enter the cell by endocytosis [5, 6]. The resulting class II:peptide complexes, products of the endosomal/lysosomal processing pathway, are recognized by CD4+ T cells. CD4+ cells, in large part, function to enhance antibody production by B lymphocytes and to activate cellular immune responses. They have thus been termed helper T cells. In contrast, MHC class I molecules primarily present peptides derived from intracellular proteins with access to proteolytic machinery within the cytosol. Resulting class I:peptide complexes are important in directing the immune response of CD8+ cytolytic T lymphocytes against intracellular pathogens such as viruses.

Much has been learned in recent years regarding the intracellular events regulating MHC-restricted presentation of antigen. Although the specifics of class I antigen processing are not yet fully understood, evidence suggests that a multisubunit proteolytic complex within the cytosol may play a role in generating antigenic peptides for class I presentation [reviewed in ref. 7]. The resulting peptides are thought to be imported into an early exocytic compartment, presumably the endoplasmic reticulum (ER) by two MHC-encoded proteins (TAP-1 and TAP-2) which belong to a family of transporter proteins known as the ABC (ATP-binding cassette) superfamily [8–11]. Peptide binding to class I molecules allows for stable association with a second non-MHC-encoded protein, β_2-microglobulin [12, 13]. The assembly of the MHC class I heavy chain with β_2-microglobulin and peptide within the ER is facilitated by an 88-kD protein named calnexin [14, 15]. Only after binding to both peptide and β_2-microglobulin does the MHC class I molecule achieve a conformation suitable for exit from the ER/cis Golgi [16, 17].

Early studies investigating MHC class II-restricted antigen presentation revealed several general characteristics. It is time-dependent and blocked by agents known to inhibit the activity of lysosomal proteases. Analysis of the intracellular trafficking of the class II molecule has also implicated the importance of endosomal compartments in class II-restricted antigen presentation. During biosynthesis, MHC class II molecules, unlike MHC class I, do not appear to follow the default ER/Golgi/cell surface exocytic pathway. While transmembrane proteins within the default exocytic pathway appear at the cell surface within minutes after sialic acid addition in the trans-Golgi, expression of MHC class II molecules occurs hours after sialylation [18]. MHC class II molecules are sensitive to endocytosed neuraminidase during this time, suggesting that these proteins intersect endocytic compartments after transport through the Golgi [19]. Interestingly, very recent subcellular fractionation studies have suggested that the class II-containing compartments may be distinct from the well-defined endosomal and lysosomal compartments [20–22]. Within these compartments or other protease containing-compartments within the endocytic pathway the invariant chain glycoprotein is released from MHC class II, an event that allows the peptide binding site within class II to become available for peptide binding [23, 24]. Thus, the currently accepted model for class II presentation of exogenous antigens involves selective trafficking of MHC class II molecules and antigen into the compartments of the endocytic pathway. Within these compartments, through the activity of resident lysosomal proteases, antigen is cleaved into peptides and invariant chain is released from the class II molecules. Peptides binding then takes place and the peptide:MHC class II complex is exported to the cell surface for recognition by CD4 T cells.

Although the class II molecule is specialized to sample extracellular antigens which enter the cell by endocytosis, it has become increasingly clear that class II molecules can also efficiently bind and present peptide-derived from proteins synthesized within the antigen-presenting cell (APC) [25–33]. Although the extent to which class II molecules display peptides derived from internally synthesized proteins is becoming increasingly apparent, the mechanisms governing their presentation remain poorly defined. In this review, I will discuss research which has provided insight into the mechanisms involved in endogenous antigen presentation and the immunological implications of endogenous antigen presentation by the MHC class II molecule.

Peptides Isolated from the MHC Class II Molecule

Recently, highly sophisticated and sensitive biochemical methods have been developed to derive sequence information from complex mixtures of peptides. Two general methods of peptide sequencing have been applied to the study of MHC-bound peptides. Both methods typically involve isolation of the MHC molecule through monoclonal antibody affinity chromatography followed by acid elution of peptides from the intact MHC molecule. Low molecular weight peptides are then separated from the larger proteins and polypeptides and the peptides are fractionated by high performance liquid chromatography (HPLC). Amino acid sequencing is then performed on individual HPLC fractions. In one method sequencing is accomplished through the use of typical Edman degradation techniques, while in the other method, sequence information is derived from tandem mass spectrometry. This latter method has the advantage of requiring less material and also has the capability to sequence individual peptides contained within complex mixtures. Both methods have provided a great deal of information on the nature of peptides naturally bound to the MHC class II molecule.

One of the early and surprising results from the analyses of peptides eluted from class II MHC molecules is that a significant fraction of the peptides isolated from MHC expressed on cells in culture is derived from internally synthesized proteins [34, 35]. Depending on the allelic product of class II molecules studied, peptides derived from internally synthetized antigens constitute from 70 to 90% of the peptides sequenced. These results suggest first that in the absence of receptor-mediated uptake of extracellular protein, internalization and/or functional utilization of exogenous proteins may be quite inefficient [36, 37], leaving the majority of class II molecules free to bind peptides derived from internally synthesized proteins. Second, these results point out that although class II molecules are indeed specialized to present antigens from an exogenous source, they are quite capable of displaying peptides that are synthesized within the APC and in fact, in the absence of pathogenic challenge, the vast majority of peptides displayed on the cell surface of an APC likely represent the array of proteins synthesized within that APC. The quantity of peptide recovered from class II molecules is close to equimolar to the MHC class II molecule itself, suggesting that the vast majority of MHC molecules are not 'empty' but rather bound by autologous peptides [38]. This finding is in agreement with earlier studies in which isolated class II was tested for the ability to bind peptide in vitro, where it was found that only a very small fraction (5–10%) of the class II molecules were active in the binding assay. Presumably, the remaining class II molecules in these studies persisted in their association with self peptides through the isolation procedure, and were thus unavailable for binding of the test peptide. Because class II molecules isolated under these procedures were eluted from anti-class II monoclonal antibody affinity columns through the use of denaturants, the inference is that once peptide is bound to the class II molecule, it remains almost irreversibly associated, in agreement with the studies made with class II molecules tested with model antigens and peptides [39–43].

With regard to structure, peptides eluted from the class II molecule are heterogeneous in size, ranging from 13 to 25 amino acids, with the majority in the range of 14–18 amino acids [34, 44–46]. Heterogeneity in the length of peptides bound to class II molecules is not only found among peptides derived from different source proteins, but also individual peptides derived from the same source peptides. Thus, the peptides bound to class II molecules are nested sets derived from a variety of cellular proteins. Heterogeneity in peptides bound to class II from a given source protein is most commonly displayed at the carboxy-terminus although amino-terminal variability is not unusual. Microheterogeneity in the termini of peptides bound the MHC class II molecule infers the activity of exopeptidases and also of flexibility in the ability of class II molecules to accommodate peptides of different sizes. This latter conclusion is supported by in vitro binding studies that have shown a great deal of permissiveness in length of the peptides that have potent binding activity to the MHC class II molecules [44]. The size microheterogeneity in peptides bound to class II is in sharp contrast to the peptides isolated from MHC class I molecules, which are almost exclusively 8–10 amino acids in length, most commonly nonomers [reviewed in ref. 47]. The longer length of peptides bound to class II despite the structural homologies in the peptide binding site in class I and class II molecules suggests that peptides in the class II binding site likely extend out from the pocket marked by the termini of the α helices.

Perhaps one of the most intriguing findings derived from elution and sequence of MHC class II-bound peptides is that an extraordinarily high percentage of source proteins are MHC molecules themselves. One very extensive study performed by Chicz et al. [45, 46] described peptides bound by six different human DR allelic products isolated from lymphoblastoid cell lines. Depending on the allele studied, MHC-related peptides constitute from 10 to 60% of the total peptides sequenced. The peptides derived from MHC molecules do not represent one particular region of the MHC molecule. Some of the peptides are derived from the helical regions in the membrane distal domains, while others were contained within the membrane-proximal external domain of either class I or class II. Preferential binding of MHC-derived peptides cannot be attributed solely to selective localization of these molecules within the endocytic pathway, as a high percentage of source proteins are derived from MHC class I molecules, which should not have preferential access to the MHC class II antigen processing compartment. There are at least two potential explanations for the predilection of MHC peptides to be present on the MHC class II molecule. First, it is possible that, contrary to conclusions from biochemical and histochemical studies, both MHC class I and class II molecules are in fact enriched in the classical endocytic compartments relative to other membrane proteins. It is conceivable that even though the intact class I molecules are not readily detected there, perhaps because of their protease sensitivity, they traffick to these compartments by additional mechanisms than those used to degrade typical cell surface proteins. Alternatively, because both MHC class I and class II molecules assemble as multicomponent complexes within the endoplasmic reticulum and because this assembly is time-dependent and limits export from the endoplasmic reticulum, a relatively high percentage of newly synthesized class I and class II molecules may actually be degraded because of their inefficient egress. Potential mechanisms for degradation of ER-localized proteins and subsequent binding to MHC class II molecules are discussed below.

Regardless of the mechanism that leads to preferential presentation of MHC-derived peptides on MHC class II molecules, the consequences of this phenomenon are interesting to consider. MHC molecules are the most genetically polymorphic molecules that exist. Based on the stability of the interaction and the evidence that peptide actually influences the conformation of the presenting molecule [48–50], peptide that is bound to an MHC molecule can be considered to be an actual component of the MHC molecule. Both MHC class I and MHC class II molecules can thus be considered to be trimeric structures. The prevalence of genetically polymorphic peptides within the peptide binding site therefore increases the diversity of the expressed MHC molecules. Thus, a given MHC molecule that is expressed on cells of divergent MHC backgrounds is essentially unique, due to the presence of divergent MHC-derived peptides within its peptide binding site. This model has particular relevance when one considers the roles of MHC polymorphism and of peptide in T cell development and in allorecognition.

Intracellular Events Involved in Presentation of Internally Synthesized Antigens by MHC Class II Molecules

Studies with Model Antigens

The intracellular mechanisms that are involved in endogenous antigen presentation by the MHC class II molecule have been probed by a number of investigatiors using model antigens. Much of the focus of these experiments has been to identifiy whether endogenous antigen presentation by class II molecules, like exogenous antigen presentation, proceeds via the conventional endosomal pathway, whether it occurs by a class I-like pathway or whether it represents a third pathway of presentation. These models have also been used to clarify the constraints of endogenous antigen presentation by class II molecules, in order to determine if all cellular antigens have equal access to the class II molecule or if presentation is restricted to proteins having identifiable characteristics or particular subcellular localizations.

Analysis on the subcellular sites of localization that allow internally synthetized antigens to gain access to class II has led to some general conclusions. First, with one exception derived using virus-infected cells [6], it has generally been found that proteins that are expressed at the cell surface can be presented by MHC class II molecules [25, 28, 30, 31, 33, 51, 52]. This has been found to be the case even when an ordinarily secreted exogenous protein (hen egg lysozyme) has been genetically manipulated so that the protein encoding the antigenic peptide was covalently linked to the transmembrane and intracytoplasmic segment of an MHC class I molecule [30]. Cell surface localization of an internally derived antigen would allow theoretical access to most sites of intracellular degradation, including that that takes place in the ER (for the newly synthetized protein), cytoplasm (for that fraction of protein translated on ribosomes that have not docked onto the ER and therefore remain in the cytoplasm) and for degradation that takes place after expression at the cell surface due to internalilzation for ultimate lysosome-mediated protein turnover. Thus, it seem likely that the majority of cell surface proteins would have the capability to provide peptides for presentation by class II molecules, provided that they contain peptides that fit the binding requirements of the particular class II molecules expressed by that individual.

In general, it has also been found that secreted proteins also have ready access to class II molecules. The options available for proteolysis are the same as that for cell surface proteins, except that access to lysosomes would be anticipated to most readily occur after secretion and reuptake. Interestingly, however, when this mechanism has been tested for the model secreted antigens that sensitize APC for recognition by CD4 T cells, it has generally been found that the mechanism used for presentation does not appear to involve secretion and reuptake [30, 33, 52]. This issue is typically addressed either by testing the ability of culture supernatant from the cell secreting the model antigen to sensitize APC for recognition, or more vigorously, by coculturing cells that secrete the antigen in question but lack the class II restriction element with cells that express the class II restriction element but lack expression of the antigen. Coculture is generally allowed to proceed for several days to mimic the conditions available to the APC that expresses both molecules. For proteins that are secreted at modest levels typical of normal cellular proteins (in the range of nanograms per milliliter), and that do not have specialized mechanisms of reuptake, secretion and endocytosis does not appear to be a viable mechanism for presentation. Nonetheless, these secreted molecules do get presented by class II molecules, raising the question of the mechanism and sites of their protein degradation and association with the class II molecule (discussed below).

To examine whether export from early biosynthetic compartments is necessary for presentation via the MHC class II molecule, as might be expected for an endosomal processing event, several strategies have been applied. For soluble proteins such as lysozyme or immunoglobulin, ER/cis Golgi retention has been accomplished by engineering an ER retention signal into the protein, encoded for by the amino acid sequence KDEL [26, 30]. To retain a transmembrane protein within the ER, our laboratory has engineered an adenovirus E19 sequence into the cytosolic tail of the model antigen we have studied (H-2L^d) [53]. Despite successful retention of the antigens tested within the ER, presentation via the class II molecules in all cases was still observed.

The site of intracellular localization that has been the most controversial is the cytoplasm. This site of expression is presumed to be the main source for peptides destined to be presented by the classical class I presentation pathway. When target antigens presented by MHC class II molecules have been localized to the cytoplasm, contrasting results have been obtained. One of the early models for endogenous antigen presentation by MHC class II molecules involved the normally cytoplasmic matrix protein of influenza virus. The matrix protein, encoded for by a recombinant vaccinia virus, was able to be presented by class II molecules, although at less efficiency than presented by the MHC class I molecule [32]. When proteins normally expressed at different subcellular localizations have been redirected to the cytoplasm (most typically achieved for normally secreted or cell membrane-expressed proteins by elimination of the signal peptide, which will prevent translocation across the ER membrane), many investigators have failed to find presentation via the class II molecule [26, 33, 54]. Some investigators have seen presentation of cytosolic-localized antigen, but this typically required high levels of protein expression [55]. Thus, it is likely that cytosolically localized antigens do have some access to the class II molecule, but presentation of these types of antigen may require atypically high levels of protein synthesis or ongoing viral infection to be presented by the class II molecule.

Mechanisms of Presentation of Internally Synthesized Antigens by MHC Class II Molecules

Based both on the use of model antigens and on knowledge of peptides naturally bound by the class II molecules, it is now generally accepted that many internally synthesized antigens can be presented by the MHC class II molcule. What is still uncertain, however, is the mechanism(s) involved in this presentation event. The primary mechanistic issues to be resolved are the sites and proteases involved in protein degradation and the sites involved in the final binding of the derived peptide to the MHC class II molecule. Potential sites for protein degradation include the endosomal/lysosomal compartments used to degrade exogenous antigens, the cytosolic proteosome system presumed to be the major pathway used by antigens destined to be presented by class I molecules [56, 57], or proteases localized to the ER that are involved in degradation of incompletely assembled or misfolded newly synthesized proteins [58, 59]. The potential sites of peptide binding to class II molecules would be the early sites of class II biosynthesis (ER/cis Golgi) and would be presumed to be analogous to MHC class I, or the endocytic compartments that are the likely site of binding of peptides derived from exogenous antigens.

Experimentally, it is difficult to distinguish among the preceding potential sites of protein degradation. The main limitation for these studies is a paucity of effective protease inhibitors that specifically inhibit the activity within one but not the other sites. For this reason, the major tool used to dissect the intracellular events of MHC-restricted antigen presentation has been the use of metabolic inhibitors. Early studies that first delineated differences between MHC class I- and MHC class II-restricted antigen presentation noted that two types of reagents appeared to selectively block the classical (endogenous) class I and classical (exogenous) class II pathways of presentation. Pioneering work on the mechanisms involved in presentation of exogenous antigens by MHC class II molecules showed that presentation was effectively blocked by treatment of APC with lysosomotropic amines such as ammonium chloride and chloroquine, implicating acid compartments in this presentation [60]. Later studies probing the events involved in MHC class I presentation utilized a fungal metabolite, brefeldin A (BFA), which blocks egress of newly synthesized proteins from the ER/cis Golgi [61]. Sensitivity of the class I presentation pathway to BFA implied that newly synthesized class I molecules were those involved in presentation of the internally synthesized antigen, a finding now consistent with data showing that class I presentation involves uptake of cytosolically degraded antigens into the ER for binding to newly synthesized class I molecules. Thus, BFA and lysosomotropic reagents have been regarded in many ways as the 'fingerprint' of a class I or class II pathway or antigen presentation, respectively.

Utilization of these metabolic inhibitors to probe class II-restricted presentation of endogenous antigens is problematic for several reasons. First, BFA is a ubiquitous inhibitor of protein export, due to its generalized effect on the organization and integrity of the ER, cis Golgi and later Golgi compartments [62, 63]. Sensitivity of an antigen presentation event to BFA thus merely implicates any newly synthesized membrane associated protein in the antigen presentation event. There is increasing evidence that both the class I and class II presentation pathways primarily involve newly synthesized MHC restriction elements, rather than preexisting molecules on the plasma membrane. Thus it is likely that most class II-restricted presentation should be BFA-sensitive, regardless of the site of peptide generation or class II binding. The longer lag time between synthesis and cell surface expression of class II molecules compared to class I molecules [18] has been used BFA studies to distinguish a class I vs. class II type presentation event. Rapid inhibition of antigen presentation by BFA is most consistent with peptide binding that takes place early in biosynthesis, followed by rapid export and cell surface display of the immunogenic MHC:peptide complex.

Lysosomotropic amines such as ammonium chloride and chloroquine have been used to implicate a role for endosomes or lysosomes in the presentation event. These studies are complicated on several accounts. First, both peptide generation (by the lysosomal proteases) and peptide binding to class II molecules occur most efficiently at acid pH [64–66]. Therefore, raising the intravesicular pH will diminish both processes. Thus, sensitivity to lysosomal amines does not assign the site of peptide generation. Secondly, degradation of invariant chain from the class II molecule is mediated by lysosomal proteases. This degradation of invariant chain is required for two key events in class II-restricted presentation: peptide binding to the class II molecule [24, 67] and export from the endosomal compartments to the cell surface [68, 69]. Our laboratory has shown that invariant chain has a potent endosomal retention function [69]. Persistent invariant chain association with class II molecules, induced by lysosomal amines, leads to retention of the class II/invariant chain complex within the endosomal compartments. Finally, many intracellular trafficking events, particularly those requiring membrane fusion, are pH-dependent and are thus inhibited by lysosomal amines [70, 71]. Thus, sensitivity of an antigen presentation event to lysosomotropic reagents most simply indicates endosomal compartments in the presentation event. These compartments could be used for pH-dependent peptide generation, peptide binding to class II or pH-dependent trafficking events. Total insensitivity to lysosomotropic reagents implies a pathway that does not involve those compartments, such as the ER-to-Golgi-to-cell surface default pathway used by class I molecules and most cell surface proteins.

Given these limitations to the utilization of these reagents, I will now discuss the results obtained when these types of experiments have been performed to dissect the events involved in class II-restricted presentation of internally synthesized antigens. One of the early studies on endogenous antigen presentation by MHC class II molecules found that presentation of internally synthesized influenza matrix protein, an exclusively cytosolic antigen, was insensitive to chloroquine but was acutely sensitive to BFA [27]. When the viral antigens were provided as an exogenous antigen, the opposite finding was made: presentatioin was sensitive to chloroquine but resistant to the effects of BFA. This suggested that two distinctive pathways could be used for presentation of the same antigen. A similar BFA sensitivity on endogenous antigen presentation was seen for an constitutively expressed class I molecule (HLA-B7) whose peptide is presented in the context of class II (DR1) [31]. This presentation was resistant to chloroquine.

Several studies have shown that internally synthesized antigens are presented by a pathway that is sensitive to lysosomotropic amines. When recombinant vaccinia virus was used to direct expression of heterologous viral proteins within human B cell lines, two cytoplasmic viral antigens (hemagglutinin matrix and a leader-minus form of the hemagglutinin), a cell surface viral protein (wild-type hemagglutinin) and even a mini-gene encoding only the peptide epitope that is synthesized within the cytoplasm were found to be presented by a chloroquine-sensitive pathway [72] within lymphoblastoid cells, as was presentation of cytosolic or ER-retained forms of hen egg lysozyme [55] within murine B cell lines. Insensitivity to lysosomotropic amines was found when measles virus proteins but not influenza proteins were presented by B cell lines or by fibroblasts [73], suggesting that sensitivity to these amines may be dependent on the mode of virus entry or to the epitope studied.

Several conclusions can be drawn from these studies. First, Malnati et al. [72] found that presentation of the viral epitope encoded by a mini-gene was sensitive to chloroquine, which is consistent with the view that chloroquine sensitivity does not identify the site of antigen processing, as the mini-gene is genetically 'preprocessed' into an immunogenic peptide. Chloroquine effects must under these conditions be due to the block in invariant chain dissociation or to disruption of endosomal trafficking events. It would be particularly informative to know whether the chloroquine effect was dependent on expression of invariant chain within the APC. Our laboratory has shown that deposition of newly synthesized class II to the plasma membrane is not blocked by lysosomotropic amines if the class II molecules are expressd in the absence of invariant chain [69], despite the fact that these molecules can be shown to traverse the endocytic pathway during biosynthesis [74]. Dependence of the chloroquine effect on invariant chain expression would implicate invariant chain dissociation as the site of chloroquine inhibition. Such a mechanism would necessitate an exploration of the mechanism by which the cytosolic peptide gains access to the endosomal compartments for binding to the MHC class II molecule.

Secondly, the finding by several groups that presentation of cytoplasmically synthesized protein is sensitive to lysosomotropic amines argues against utilization of a classical class I presentation pathway of cytoplasmic protein degradation and endoplasmic reticulum binding. Because peptide binding to the class II molecule has been shown to be blocked by the presence of invariant chain, and because invariant chain has been implicated in sorting to endosomal compartments, it has been predicted [75] that stable binding of peptide to class II molecules in the ER, as would occur in a class I-like antigen processing pathway, might lead to trafficking of class II molecules by the default pathway to the cell surface. If such is the case, two phenotypes should be observed as a consequence of ER-derived peptide binding to class II molecules: independence from chloroquine, whose effects would only be detectable in endocytic events, and acute sensitivity to the effects of BFA. The fact that cytosolic antigens are inhibited by chloroquine thus implicates endosomal compartments in some aspect of the presentation event, which tends to argue against binding of the peptide in the ER, assuming the model discussed above on the consequences of early loading of class II is correct. Lack of utilization of the classical class I pathway for cytosolic antigens presented by class II molecules is also supported by the observation that, when studied, presentation of cytoplasmic antigens by the class II molecule [72], like presentation of other internal antigens by class II molecules [33], is independent of TAP gene expression, thought to be involved in import of cytoplasmically generated peptides into the ER for MHC class I binding [8, 9, 11, 76].

Thus, at the present time, a major unresolved question related to class II-restricted presentation is whether the endosomal/lysosomal compartments are the exclusive site of peptide binding to class I molecules, to be used regardless of the intracellular site of peptide generation. The endosomal compartments have been implicated as the major site of peptide class II interactions based on the ability of low pH to facilitate peptide binding to class II [64–66] and through the observation that, upon entry into endosomal compartments and release of the invariant chain, MHC class II molecules obtain a conformational state that is associated with stable peptide binding (maintenance of $\alpha\beta$ dimer formation in SDS) [77, 78]. It is not yet clear how cytosolic or ER-localized antigens would gain access to the MHC class II molecule, especially if binding requires access to the endosomal compartments. There is some evidence that cytosolic and ER proteins can enter the classical lysosomal/endosomal compartments through the processes of autophagy and chaperone-mediated uptake [79–81], processes induced by cellular stresses such as nutritional deprivation and viral infection. Autophagy is thought to involve formation of a cytoplasmic vacuole from the ER membranes (the autophagic vacuole). This vacuole thus likely contains resident ER proteins and also soluble cytosolic proteins. The vacuole is then thought to completely separate from the ER and eventually fuse with lysosomal membranes, thus delivering its contents to this latter organelle. In contrast, heat shock protein-mediated uptake is thought to involve direct translocation of cytosolic protein into the lysosome. This process involves selective recognition of particular amino sequences (related to KFERQ) and thus is thought to be selective for particular proteins.

In addition to providing a mechanism to allow localization of ER and cytosolic antigens to the lysosome for binding to MHC class II molecules, there is another interesting consequence that would derive from such processes. If particular metabolic and/or pathogenic events increase that access of ER and cytosolic proteins to the MHC class II molecule by these two intracellular trafficking pathways, then the array of self as well as foreign peptides bound to class II molecules will change under these conditions. These processes thus might reveal new peptide:class II complexes at the cell surface. Even 'self' peptides, having never previously had access to class II molecules, might thus be regarded as 'foreign' by the immune systen and initiate an autoimmune response.

Role of Invariant Chain in the Presentation of Endogenous Antigens by the MHC Class II Molecule

The invariant chain glycoprotein assembles with the MHC class II αβ dimer soon after biosynthesis in the ER. A number of biological activities have been demonstrated for invariant chain that influences antigen presentation by the MHC class II molecule [reviewed in ref. 82]. Two of these activities would be particularly expected to influence endogenous antigen presentation by class II molecules. The first is the ability of invariant chain to antagonize peptide binding to the MHC class II molecule. This property of invariant chain has now been demonstrated by several investigators [24, 67, 75]. The ability of invariant chain to block peptide bindings has been interpreted as the mechanism that would prevent internal antigens from binding to class II molecules early in biosynthesis and thus would distinguish the site of class I antigen binding from that involved in class II peptide binding. By this model, assembly of invariant chain with the class II molecule early in biosynthesis would preclude binding of ER-available peptides, such as those that are presented by MHC class I molecules. The peptide binding site for class II would thus only be available after invariant chain is proteolytically released from the class II molecule in endosomal compartments. If this model is correct, any early peptide binding to class II molecules would be expected to be antagonized by the presence of invariant chain within the APC.

The second activity of invariant chain that would influence class II-restricted presentation of internally synthesized antigens is its ability to localize the class II molecule within endocytic compartments, a property controlled by the cytoplasmic tail of invariant chain. The amino-terminal cytosolic segment of invariant chain is one of the last proteolytic fragments to be released from the class II molecule. This segment thus might retain class II molecules in endocytic compartments after the peptide binding site has been freed of invariant chain and is made competent for peptide binding. Therefore, internal antigens that require endosomal loading of class II or long steady-state expression of the class II molecule within these compartments might be expected to be enhanced by expression of invariant chain within the APC.

Until recently, it has been difficult to demonstrate a negative influence of invariant chain on MHC class II-restricted presentation of endogenous antigens. When examined in several model systems, invariant chain has been shown to either have no effect [33] or to facilitate endogenous antigen presentation by class II molecules [83, 84], apparently arguing against one of the earliest models for invariant chain function: to block peptide binding to class II molecules in the ER. Rather, the studies showing a positive influence of invariant chain on endogenous antigen presentation argue in favor of the endosomal compartments as the site of peptide:class II molecule complex formation, even for internally synthesized antigens. However, several recent reports [84, 85] have shown that for some antigens, invariant chain expression within the APC can antagonize class II-restricted presentation. The simplest interpretation of these recent experiments is that those class II:peptide epitopes blocked by invariant chain are formed in the ER. The potential sites of antigen proteolysis would then be presumed to be either the ER or the cytosol. Interestingly, in the studies of Bodmer et al. [84], it was shown that APC lacking expression of invariant chain express unique class II:peptide epitopes that are recognized as foreign by T cells from normal mice. This finding suggests the possibility that when cells in the body are induced to become class II-positive, for instance during an inflammatory response, but do not coinduce invariant chain, they may be able to initiate an autoimmune response, due to their display of class II:peptide complexes that the circulating T cell pool is not tolerant to. Alternatively, they may be particularly efficient at presenting internally synthesized antigens such as tumor antigens, that are expressed primarily in early biosynthetic compartments or in the cytosol.

Concluding Remarks

It is now clear that a significant fraction of peptides that are displayed by the MHC class II molecule are derived from proteins synthesized within the APC itself. The proteins that contribute to the class II peptide repertoire are likely to be those that enter the exocytic pathway and are destined to be expressed ultimately as a secreted or membrane-associated form. Nonetheless, it is also clear that even cytosolic proteins can gain access to the class II molecule, particularly in cases when they are expressed at high levels or during viral infection. Thus it is likely that the capacity of class II molecules to display peptides from pathogens or altered self proteins that are uniquely expressed in tumor cells is quite diverse, which is of course of benefit to the host.

What is less clear at the present time are the intracellular events involved in class II-restricted presentation of endogenous antigens. There remain several major issues to be resolved. The first is the potential for the ER to serve as a site of peptide generation and peptide binding to the class II molecule. Although there are now several findings that argue that this can occur, it is not yet clear whether these represent rare exceptions or if the ER is a reasonably viable option for class II molecules to associate with peptide. Aside from its importance from the point of view of the cell biology of class II molecules, one of the main consequences of peptide binding to class II molecules within the ER is that there is the potential for invariant chain expression to impact on the array of peptides presented by class II molecules. Lack of invariant chain would lead to the association of a distinctive set of peptides with the class II molecules, some of which may have the potential to serve as a target of an autoimmune response, while others may provide targets for protective immune responses to intracellular pathogens or tumors.

A second major unresolved issue is the intracellular trafficking pattern that allows cytosolic and ER-retained antigens to gain access to the class II molecule. Of particular interest are the possibilities that these proteins may gain access to endosomal compartments by such processes as autophagy or heat shock protein-mediated uptake. Both of these processes may be selective for certain proteins and might be regulated by the metabolic status of the host cell and by conditions of stress or pathogenic insult. Thus, if these mechanisms do control import of cytosolic antigens and/or ER-derived peptides into the class II molecule binding compartment, the display of peptides derived from self as well as foreign proteins on the class II molecule reflects dynamic rather than static processes. Similarly, the efficiency and conditions that induce these processes may vary from one cell type to another, leading to cell type-dependent variability in the display of self as well as foreign proteins by the MHC class II molecule. Such heterogeneity could have several potential consequences. It might lead to selective participation of certain APC in the response to pathogens, which in turn has the potential to regulate the T cell effector function elicited. In addition, dissimilarity in the array of self peptides displayed by distinctive cell types would likely influence self tolerance induction and the potential for autoimmunity.

References

1 Bjorkman P, Saper M, Samraoui B, Bennet W, Strominger J, Wiley D: The foreign antigen binding site and T cell recognition regions of class I histocompatibilty antigens. Nature 1987;329:512–518.

2 Blackwell J, Roberts M: Immunomodulation of murine visceral leishmaniasis by administration of monoclonal anti-I-A antibodies: Differential effects of anti-I-A vs anti-I-E antibodies. Eur J Immunol 1987;17:1669–1672.

3 Stern L, Wiley D: The human class II MHC protein HLA-DR1 assembles as empty αβ heterodimers in the absence of antigenic peptide. Cell 1992;68:465–477.

4 Brown J, Jardetzky T, Goga J, Stern L, Urban R, Strominger J, Wiley D: Three-dimensional structure of the human class II histocompatibility antigen HLA-DR1. Nature 1993; 364:33–39.

5 Germain RN: The ins and outs of antigen processing and presentation. Nature (London) 1986;322:687–688.

6 Morrison L, Lukacher A, Braciale V, Fan D, Braciale T: Differences in antigen presentation to MHC class I- and class II-restricted influenza virus-specific cytolytic T lymphocyte clones. J Exp Med 1986;163: 903–921.

7 Monaco J: A molecular model of MHC class I-restricted antigen processing. Immunol Today 1992;13: 173–179.

8 Attaya M, Jameson S, Martinez C, Hermel E, Aldrich C, Forman J, Lindahl K, Bevan M, Monaco J: Ham-2 corrects the class I antigen processing defect in RMA-S cells. Nature (London) 1992;355:647–649.

9 Deverson E, Gow I, Coadwell W, Monaco J, Butcher G, Howard J: MHC class II region encoding proteins related to the multidrug resistance family of transmembrane transporters. Nature 1990;348:738–741.

10 Spies T, Bresnahan M, Bahram S, Arnold D, Blanck G, Mellins E, Pious D, DeMars R: A gene in the human major histocompatibility complex class II region controlling the class I antigen presentation pathway. Nature 1990;348:744–747.

11 Powis S, Townsend A, Deverson E, Bastin J, Butcher G, Howard J: Restoration of antigen presentation to the mutant cell line RMA-S by an MHC-linked transporter. Nature 1991;354:528–531.

12 Kozlowski S, Takeshita T, Boencke W.-H., Takahashi H, Boyd L, Germain R, Berzofsky J, Margulies D: Excess β2 microglobulin promoting functional peptide association with purified soluble class I MHC molecules. Nature 1991;349:74–77.

13 Ribaudo R, Margulies D: Independent and synergistic effects of disulfide bond formation, beta 2 microglobulin, and peptides on class I MHC folding and assembly in an in vitro translation system. J Immunol 1992;149:2935–2944.

14 Degen E, Williams D: Participation of a novel 88-kD protein in the biogenesis of murine class I histocompatibility molecules. J Cell Biol 1991;112:1099–1115.

15 Ahluvalia N, Bergeron JJM, Wada I, Degan E, Williams D: The p88 molecular chaperone is identical to the endoplasmic reticulum membrane protein, calnexin. J Biol Chem 1992; 267:10914–10918.

16 Townsend A, Ohlen C, Bastin J, Ljunggren J, Foster L, Karrer K: Associaton of class I major histocompatibility heavy and light chains induced by viral peptides. Nature 1989;340:443–448.

17 Ljunggren H, Stam N, Ohlen C, Neefjes J, Hoglund P, Heemels M, Bastin J, Schumacher T, Townsend A, Karre K, Ploegh H: Empty MHC class I molecules come out in the cold. Nature 1990;346:476–480.

18 Neefjes J, Stollorz V, Peters P, Geuze H, Ploegh H: The biosynthetic pathway of MHC class II but not class I molecules intersects the endocytic route. Cell 1990;61:171–183.

19 Cresswell P: Intracellular class II HLA antigens are accessible to transferrin-neuraminidase conjugates internalized by receptor mediated endocytosis. Proc Natl Acad Sci USA 1985;82:8188–8192.

20 Amigorena S, Drake J, Webster P, Mellman I: Transient accumulation of new class II MHC molecules in a novel endocytic compartment in B lymphocytes. Nature 1994;369: 113–126.

21 Qiu Y, Xu X, Wandinger-Ness A, Dalke DP, Pierce SK: Separation of subcellular compartments containing distinct functional forms of MHC class II. J Cell Biol 1994;125: 595–605.

22 Tulp A, Verwoerd D, Dobberstein B, Ploegh H.B, Pieters J: Isolation and characterization of the intracellular MHC class II compartment. Nature 1994;369:120–126.

23 Roche P, Cresswell P: Proteolysis of the class II-associated invariant chain generates a peptide binding site in intracellular HLA-DR molecules. Proc Natl Acad Sci USA 1991; 88:3150–3154.

24 Roche P, Cresswell P: Invariant chain association with HLA-DR molecules inhibits immunogenic peptide binding. Nature 1990;345: 615–618.

25 Weiss S, Bogen B: B lymphoma cells process and present their endogenous Ig to MHC restricted T cells. Proc Natl Acad Sci USA 1989;86: 282–286.

26 Weiss S, Boen B: MHC class II restricted presentation of intracellular antigen. Cell 1991;64:767–776.

27 Nuchtern J, Biddison W, Klausner R: Class II MHC molecules can use the endogenous pathway of antigen presentation. Nature 1990;343:74–76.

28 Bikoff EK: Class II restricted IgG2a^{b}-specific T cells recognize a signal minus form of the V-CH3^{b} antigen. Eur J Immunol 1991;21: 1411–1417.

29 Bikoff E: Formation of complexes between self-peptides and MHC class II molecules in cells defective for presentation of exogenous proteins antigens. J Immunol 1992;149: 1–8.

30 Brooks A, Hartley S, Kjer-Nielsen L, Perara J, Goodnow C, Basten A, McCluskey J: Class II-restricted presentation of an endogenously derived immunodominant T cell determinant of hen egg lysozyme. Proc Natl Acad Sci USA 1991;88:3290–3294.
31 Chen B, Madrigal A, Parham P: Cytotoxic T cell recognition of an endogenous class I HLA peptide presented by a class II HLA molecule. J Exp Med 1990;172:779–788.
32 Jaraquemad D, Marti M Long EO: An endogenous processing pathway in vaccinia virus-infected cells for presentation of cytoplasmic antigens to class II-restricted T cells. J Exp Med 1990;172:947–954.
33 Loss GE, Elias CE, Fields PE, Ribaudo RK, McKisic M, Sant AJ: MHC class II-restricted presentation of an internally synthesized antigen displays cell-type variability and segregates from the exogenous class II and endogenous class I antigen presentation pathways. J Exp Med 1993;178:73–85.
34 Rudensky A, Preston-Hurlburt P, Hong S-C, Barlow A, Janeway CA: Sequence analysis of peptides bound to MHC class II molecules. Nature 1991;353:622–627.
35 Hunt D, Henderson R, Shabanowitz J, Sakaguchi K, Michel H, Sevilir N, Cox A, Appella E, Engelhard V: Characterization of peptides bound to the class I MHC molecule HLA-A2.1 by mass spectrometry. Science 1992;255:1261–1263.
36 Lanzavecchia A: Receptor-mediated antigen uptake and its effect on antigen presentation to class II-restricted T lymphocytes. Annu Rev Immunol 1990;8:773.
37 Castan LA, Pierce SK: Receptor mediated B cell antigen processing: Increased antigenicity of a globular protein covalently coupled to antibodies specific for B cell surface structures. J Immunol 1988; 140:404–415.
38 Buus S, Sette A, Colon S, Grey H: Autologous peptides consistutively occupy the antigen binding site on Ia. Science 1988;242:1045–1047.
39 Babbitt, B, Allen P, Matsueda G, Haber E, Unanue E: The binding of immunogenic peptides to Ia histocompatibilty molecules. Nature 1985;317:359–361.
40 Buus S, Sette A, Colon S, Jenis D, Grey H: Isolation and characterization of antigen-Ia complexes in T cell recognition. Cell 1986;47:1071–1077.
41 Watts T, McConnell H: High affinity fluorescent peptide binding to I-A^d in lipid membranes. Proc Natl Acad Sci USA 1986;83:9660–9664.
42 Lanzavecchia A, Reid P, Watts C: Irreversible association of peptides with class II MHC molecules in living cells. Nature 1992;357:249–252.
43 Jensen PE: Long-lived complexes between peptide and class II major histocompatibility complexes are formed at low pH with no requirement for pH neutralization. J Exp Med 1992;176:793–798.
44 Hunt D, Michel H, Dickinson T, Shabanowitz J, Cox A, Sakaguchi K, Appella E, Grey H, Sette A: Peptides presented to the immune system by the murine class II major histocompatibility complex molecule I-A^d. Science 1992;256:1817–1820.
45 Chicz R, Urban R, Lane W, Gorga J, Stern L, Vignali D, Strominger J: Predominant naturally processed peptides bound to HLA-DR1 are derived from MHC-related molecules and are heterogenous in size. Nature 1992;358:764–768.
46 Chicz RM, Urban RG, Gorga JG, Vignali DAA, Lane WS, Strominger JL: Specificity and promiscuity among naturally processed peptides bound to HLA-DR alleles. J Exp Med 1993;178:27–47.
47 Engelhard VH: Structure of peptides associated with MHC class I molecules. Curr Opin Immunol 1994;6: 13–23.
48 Catipovic B, Dal-Porto J, Mage M, Johansen T, Schneck J: Major histocompatibility complex conformational epitopes are peptide specific. J Exp Med 1992;176:1611–1618.
49 Bluestone J, Jameson S, Miller S, Dick R: Peptide induced conformational changes in class I heavy chains alter major histocompatibility complex recognition. J Exp Med 1992;176:1757–1761.
50 Fremont D, Matsumara M, Stura E, Peterson P, Wilson I: Crystal structures of two viral peptides in complex with murine MHC class I H-2kb. Science 192;257:919–927.
51 Thomas DB, Hodgson J, Riska PF, Graham CM: The role of the endoplasmic reticulum in antigen processing: N-glycosylation of influenza hemagglutinin abrogates CD4+ cytotoxic T cell recognition of endogenously processed antigen. J Immunol 1990;144:2789.
52 Calin–Laurens V, Forquet F, Lonmabrd-Plate S, Bertolino P, Chretine I, Trescol-Biemont M, Gerlier D, Rabourdin-Combe C: High efficiency of endogenous antigen presentation by MHC class II molecules. Int Immunol 1992;4:1113–1121.
53 Chottanapund K, Sant AJ: Early biosynthetic forms of H-$2L^d$ gain access to the MHC class II molecule. In preparation 1994.
54 Calin-Laurens V, Forquet F, Mottez E, Kanellopoulos J, Godeau F, Kourilsky P, Gerlier D, Rabourdin-Combe C: Cytosolic targeting of hen egg lysozyme gives rise to a short-lived protein presented by class I but not class II major histocompatibility complex molecules. Eur J Immunol 1992;21:761–769.
55 Brooks A, McCluskey J: Class II-restricted presentation of a hen egg lysozyme determinant from endogenous antigen sequestered in the cytoplasm or endoplasmic reticulum of the antigen presenting cells. J Immunol 1993;150:3690–3697.
56 Brown M, Driscoll J, Monaco J: Structural and serological similarity of MHC-linked LMP and proteasome (multicatalytic proteinase) complexes. Nature 1991;353:355–357.
57 Patel S, Monaco J, McDevitt H: Delineation of the subunit composition of human proteasomes using antisera against the major histocompatibility complex encoded LMP2 and LMP7 subunits. Proc Natl Acad Sci USA 1994;91:296–300.
58 Lippincott-Schwartz J, Bonifacino J, Yuan L, Klausner R: Degradation from the endoplasmic reticulum: Disposing of newly synthesized proteins. Cell 1988;54:209–220.
59 Stafford FJ, Bonfacino JS: A permeabilized cell system identifies the endoplasmic reticulum as a site of protein degradation. J Cell Biol 1991;115:1225–1236.

60 Ziegler HK, Unanue ER: Decrease in macrophage antigen catabolism caused by ammonia and chloroquine is associated with inhibition of antigen presentation to T cells. Proc Natl Acad Sci USA 1982;79:175–178.
61 Nuchtern J, Bonifacino J, Biddison W, Klausner R: Brefeldin A implicates egress from endoplasmic reticulum in class I restricted antigen presentation. Nature 1989;339:223–226.
62 Klausner RD, Donaldson JG, Lippincott-Schwartz J: Brefeldin A: Insights into the control of membrane traffic and organelle structure. J Cell Biol 1992;116:1071–1080.
63 Kreis A: Regulation of vesicular and tubular membrane traffic of the Golgi complex by coat proteins. Curr Opin Cell Biol 1992;4:609–615.
64 Harding C, Roof R, Allen P, Unanue E: Effects of pH and polysaccharides on peptide binding to class II major histocompatibility complex molecules. Proc Natl Acad Sci USA 1991;88:2740–2744.
65 Jensen P: Regulation of antigen presentation by acidic pH. J Exp Med 1990;171:1779–1784.
66 Wettstein D, Boniface J, Reay P, Schild H, Davis M: Expression of a class II major histocompatibility complex (MHC) heterodimer in a lipid-linked form with enhanced peptide/soluble MHC complex formation at low pH. J Exp Med 1991;174:219–228.
67 Teyton L, O'Sullivan D, Dickson P, Lotteau V, Sette A, Fink P, Peterson P: Invariant chain distinguishes between the exogenous and endogenous antigen presentation pathways. Nature 1990;348:39–44.
68 Neefjes J, Ploegh H: Inhibition of endosomal proteolytic activity by leupeptin blocks surface expression of MHC class II molecules and their conversion to SDS resistant αβ heterodimers in endosomes. EMBO J 1992;11:411–416.
69 Loss GE, Sant AJ: Invariant chain retains MHC class II molecules in the endocytic pathway. J Immunol 1993;150:3187–3197.
70 Seglen PO: Inhibitors of lysosomal function. Methods Enzymol 1983;96:737–765.
71 Seglen PO, Bohley P: Autophagy and other vacuolar protein degradation mechanisms. Experientia 1992;48:158–172.
72 Malnati M, Marti M, LaVaute T, Jaraquemada D, Biddison W, DeMars R, Long E: Processing pathways for presentation of cytosolic antigen to MHC class II-restricted T cells. Nature 1992;357:702–704.
73 Sekaly R, Jacobsen S, Richert J, Tonnelle C, McFarland H, Long E: Antigen presentation to HLA class II-restricted measles virus-specific T cell clones can occur in the absence of the invariant chain. Proc Natl Acad Sci USA 1988;85:1209–1213.
74 Chervonsky AC., Gordon L, Sant AJ: MHC class II β chain express a targeting signal that directs localization of class II within the endocytic pathway. Int Immunol 1994;6:973–982.
75 Bijlmaker MJ, Benaroch P, Ploegh H: Assembly of HLA-DR molecules translated in vitro: Binding of peptide in the endoplasmic reticulum precludes association with invariant chain. EMBO J 1994;13:2629–2707.
76 Trowsdale J, Hanson I, Mockridge I, Beck S, Townsend A, Kelly A: Sequence encoded in the class II region of the MHC related to the 'ABC' superfamily of transporters. Nature 1990;348:741–744.
77 Germain R, Hendrix L: MHC class II structure, occupancy and surface expression determined by post-endoplasmic reticulum antigen binding. Nature 1991;353:134–139.
78 Sadegh-Nasseri S, Germain RN: A role for peptide in determining MHC class II structure. Nature 1991;353:167–170.
79 Dunn WA: Studies on the mechanisms of autophagy: Formation of the autophagic vacuole. J Cell Biol 1990;110:1923–1935.
80 Chiang H, Terlecky S, Plant CP, Dice J: A role for a 70 kilodalton heat shock protein in lysosomal degradation of intracellular proteins. Science 1989;246:382–385.
81 Dice J: Peptide sequences that target cytosolic proteins for lysosomal proteolysis. Trends Biochem Sci 1990;15:305–309.
82 Sant AJ, Miller J: MHC class II antigen processing: Biology of invariant chain. Curr Opin Immunol 1994;6:57.
83 Humbert M, Bertolineo P, Forquet F, Rabourdin-Combe C, Gerlier D, Davoust J, Salamero J: Major histocompatibility complex class II-restricted presentation of secreted and endoplasmic reticulum resident antigens requires the invariant chains and is sensitive to lysosomotropic agents. Eur J Immunol 1993;23:3167–3172.
84 Bodmer H, Viville S, Benoist C, Mathis D: Diversity of endogenous epitopes bound to MHC class II molecules limited by invariant chain. Science 1994;2263:1284–1286.
85 Long E.O, Lavaute T, Pinet V, Jaraquemada D: Invariant chain prevents HLA-DR-restricted presentation of a cytosolic peptide. J Immunol 1994;153:1487–1494.

Immunol Res 1994;13:268–279

Roger Sciammas
Y. Tatsumi
A.I. Sperling
K. Arunan
Jeffrey A. Bluestone

Department of Pathology,
Ben May Institute and Committee
on Immunology, University of
Chicago, Ill., USA

TCRγδ Cells: Mysterious Cells of the Immune System

Key Words
TCRγδ cells
T cell receptor
Clonal selection
Antigen recognition
Immunological protection

Abstract
γδ cells participate in pathogenic infections and autoimmune conditions, yet, almost a decade after their discovery, little is known regarding their TCR repertoire or effector functions. Unlike MHC-restricted antigen recognition employed by TCRαβ cells, TCRγδ cells can recognize whole unprocessed antigens in an MHC-independent manner. The nature of positive and negative selection used to shape the repertoire of TCRγδ cells is unclear, especially in the nonlymphoid tissues where these cells predominate. While TCRγδ cells express an activated phenotype and are present in pathological conditions, their roles in immunological protection is unknown. This review will focus on our efforts to study these issues of TCRγδ biology.

Introduction

T cell receptor γδ (TCRγδ) cells continue to be an enigmatic cell subset in the immune system. TCRγδ cells are clearly dominant in certain autoimmune conditions, appearing in enhanced numbers of gut lesions of celiac sprue, polymyositis, and in the synovial fluid of joints affected by rheumatoid arthritis [1]. In addition, the numbers of TCRγδ cells increase in response to pathogenic challenge by viruses (HSV, HIV, influenza), bacteria *(Listeria monocytogenes, Mycobacterium tuberculosis)* and parasites *(Leishmania major, Schistosoma mansoni, Trypanosoma cruzi),* suggesting that these cells are important in regulating immune responses to these pathogens [1]. However, in most instances, it has been difficult, if not impossible, to identify a precise antigen specificity for these cells in these diseases. In fact, the lack of significant involvement of TCRγδ in most classical T cell responses suggests that they perform a unique and special role in host immune defense. In

Jeffrey A. Bluestone, PhD
Committee on Immunology
5841 South Maryland Avenue MC 1089
Chicago, IL 60637 (USA)

0257–277X/94/
0134–0268$8.00/0

this regard, TCRγδ cells have several unique features that distinguish them from TCRαβ cells. First, the majority of TCRγδ cells have a restricted tissue distribution [2]. TCRγδ cells predominate in the epithelial lining of the intestine, lung, skin, and reproductive tract. Interestingly, these resident TCRγδ cells have a more limited TCR repertoire than TCRαβ cells. In fact, in some epithelial tissues, a single TCRγ and δ chain with limited junctional diversity is expressed. Second, the TCRγδ cells develop in waves within the thymus prior to TCRαβ development, with each successive wave expressing a distinct family of TCR Vγ and Vδ genes [3]. The reason for this limited TCRVγVδ pairing, restricted VDJ junctional diversity, and selective tissue expression remains unclear but may reflect both molecular regulation and selection during either intra- or post-thymic development. Finally, TCRγδ cells do not appear to recognize the same types of antigens as TCRαβ cells [1, 4]. Selected antigens, such as heat shock proteins and a nonpolymorphic keratinocyte-derived antigen, are common ligands for TCRγδ cells [5, 6]. This finding has led to speculation that these cells, in general, may recognize nonpolymorphic, conserved antigens expressed during stress responses. However, we, and others, have shown that TCRγδ cells can recognize a variety of diverse antigens including: class I and class II MHC-encoded molecules; nonclassical major histocompatibility complex (MHC)-encoded molecules (CD1, Qa-1, TCT.1, TL); bacterial products other than heat shock proteins (endotoxin, and a 5′ triphosphorylated thymidine-containing compound), and a limited number of nominal antigens (G-T copolymer, herpesvirus gI and tetanus toxoid) [1]. Thus, the TCRγδ antigen repertoire and their role in immune responses may indeed be diverse.

Recent advances suggest, perhaps for the first time, that a more precise understanding of the TCRγδ population is possible. First, the advent of Cα null mice has allowed immunologists to directly investigate the ability of TCRγδ cells to protect against infections [7, 8]. These studies have demonstrated that TCRγδ do indeed possess immunoregulatory functions in the absence of TCRαβ cells implying that TCRγδ cells respond to antigenic stimuli during the course of normal immune responses. Second, recent evidence demonstrates that TCRγδ may recognize antigens in a novel fashion, similar to immunoglobulins. These results suggest that the types of antigens or antigen-presenting substrates recognized by the TCRγδ cells direct the effector functions of this unique subpopulation [4, 9]. Finally, TCRγδ transgenic mice have been developed that express TCRγδ cells of a known specificity [10, 11]. This provides a model system for the evaluation of TCRγδ cells during the development of immune responses and during selection while manipulating the context of antigen presentation. In this review we summarize recent studies from our laboratory that address several questions related to this unique population of T cells. First, how do TCRγδ cells develop and are they susceptible to positive or negative selection? Second, how do TCRγδ cells recognize nominal antigen; are they recognized as processed peptides presented in the context of a presenting molecule or as whole proteins? Finally, what is the role of TCRγδ cells in the development of disease and regulation of normal immune responses?

How Do TCRγδ Cells Develop?

Much of what we know about the development of T cells derives from the systematic study of the maturation and repertoire selection of TCRαβ cells. TCRαβ cells recognize peptide antigens presented in the context of self-MHC [12]. The process of thymic education known as positive and negative selection results in the immune system's ability to discriminate between self and nonself antigens [13]. During TCRαβ development in the thymus, immature T cells interact with self-MHC. If this interaction is productive, the T cells are signalled to mature and are thus positively selected. If the selected T cells interact too avidly with the self-MHC molecule, the T cells undergo apoptosis and die (clonal deletion/negative selection). Although the selection events governing the production of thymus-derived αβ cells are relatively well characterized, it is unclear whether TCRγδ T cells follow analogous pathways. Selection and expansion of TCRγδ cells have been demonstrated [3]. For example, TCRγδ cells that migrate to the skin and reproductive tract as intraepithelial lymphocytes appear to be selected during fetal thymus development. Although one set of studies has clearly shown that the selection of the canonical Vγ3 rearrangement observed in skin TCRγδ cells is based on a molecular mechanism, another intriguing study suggests a biological-based selection by demonstrating that in a Vγ3 transgenic mouse, a non-MHC-linked deletion of the Vγ3 TCR transgene occurred in certain mouse strains but not others [14, 15]. In another instance, a dominant δ junctional sequence is present in BALB/c lung TCRγδ cells while absent in other strains [16]. Lefrancois et al. [17] have shown that intestinal intraepithelial lymphoytes undergo positive selection based on class II, I-E^k, expression. In humans, an age-related expansion of a specific TCRγδ subset (Vδ2/Vγ9) occurs postnatally in the periphery [18].

We have generated and studied a transgenic mouse, G8, which expresses MHC-specific TCRγ and δ transgenes [10]. The G8 clone was generated from a mixed lymphocyte culture and is specific for MHC-encoded molecules, T10 and T22, expressed as alloantigens in H-2^k and H-2^b but not H-2^d mice. Thus, G8 transgenic mice bred onto the original H-2^d background mature normally and express the G8 transgenes on all TCRγδ cells independent of their location. Most importantly, these cells respond to the original MHC T region-encoded antigen, T10^b as an alloantigen. The development of mature Tg G8 T cells is dependent on positive selection following the interaction of the TCR on a self-MHC class I molecule(s) [19]. These studies used the class I-deficient, β_2-microglobulin (β_2M) null mouse produced by disruption of the β_2M gene by homologous recombination. Since previous studies had shown that recognition of T10^b by G8 depended on β_2M expression, this mouse strain was ideal for determining the role of ligand binding in selecting this TCRγδ subset. We have found that transgenic TCRγδ cells do not mature in the thymus of β_2M-deficient mice, cannot signal through their TCR, and do not exit from the thymus to populate peripheral lymphoid organs. Similarly, treatment of normal G8 mice with cyclosporine A (a drug previously shown to block positive selection of TCRαβ cells) blocks TCRγδ development in Tg as well as normal mice [20]. In addition, prior to positive selection, these immature TCRγδ cells are uniformly HSA$^+$ and continue to transcribe the recombination enzymes, RAG-1 and RAG-2, in contrast to the mature HSA$^-$ cells. These results closely parallel the general parameters of αβ T cell-positive selection, and indicate that at least some TCRγδ cells require a TCR-self antigen interaction to complete differentiation.

The molecular and biochemical signals that direct immature T cells to proliferate and differentiate after they enter the thymus are largely unknown. We have begun to directly examine the signalling and cellular requirements for TCRγδ cell maturation using fetal thymus organ cultures and in vitro cultured thymic stromal cells [Tatsumi et al., unpublished results]. In these studies, we have found that the immature population of TCRγδ cells isolated from the thymus of transgenic G8 crossed to β_2M null mice fail to proliferate, produce interleukin-2 (IL-2), or differentiate after TCR ligation despite normal early signalling events such as calcium flux; phosphorylation of the src-family kinase, lck; or activation of PLCγ1. Further analysis using gel retardation assays suggested that NF-AT binding activity to DNA in the nucleus was not evident following TCR ligation in this immature population. This contrasts with mature TCRγδ cells which exhibit increased NF-AT binding activity following receptor ligation [21]. In collaboration with Dr. Anjana Rao at Harvard University, we have found that the NF-AT activity was present in the cytoplasm of these cells and could be translocated following signalling. However, for an as yet undetermined reason, the NF-AT does not bind to the appropriate sequences of the IL-2 gene. These results suggest that important transcriptional regulatory events are modulated in the immature TCRγδ cells and it would appear that the major defect in these cells is their inability to produce IL-2 in response to TCR signalling. Interestingly, these immature TCRγδ cells will respond to antigen in vitro following the addition of IL-1 or IL-7 (two cytokines shown to be involved in T cell development and activation) to the activation assays. These data suggest that IL-1 and IL-7 may bypass the requirement for NF-AT-mediated IL-2 transcription or turn on NF-AT through a distinct signalling pathway.

The G8 transgenic mouse was used to evaluate the role of negative selection in TCRγδ development [22]. When Tg TCR^+ mice were bred to ligand-bearing, H-2^b mice, Tg TCRγδ cells were largely deleted from the thymus and spleen. This process begins during the first week of life in the thymus and is complete in the spleen by 6 weeks of age. The few cells remaining were isolated and were found to be nonresponsive to $T10^b$ stimulation. Further investigation revealed that these cells were undergoing programmed cell death. Thus, it would appear that in this model, potentially self-reactive TCRγδ cells undergo clonal deletion. However, recent studies examining TCRγδ cells that develop 'extrathymically' in the intestine suggest a distinct mechanism for maintaining tolerance in this compartment of the Tg mice [23]. We have shown that Tg^+ TCRγδ cells, with normal levels of TCR expression, are found in the intraepithelial compartments of ligand-bearing H-$2^{b/d}$ mice. The TCRγδ cells were largely unresponsive as measured by lymphokine production, induction of IL-2 receptor, and proliferation even in the presence of exogenous IL-2. Thus, the maintenance of self-tolerance of TCRγδ cells localized in epithelial tissues was due to functional unresponsiveness following antigen exposure. The differential mechanisms of tolerance could be due to differences in the potency of the antigenic stimulus in the thymus versus the epithelial tissue of the gut. Further analysis demonstrated a distinctly lower level of antigen presentation by the intestinal epithelia [24]. These results suggested that the difference between clonal deletion and clonal anergy may depend on either the avidity of the T cell/target cell interaction or the presence of costimulatory molecules on the antigen-presenting cells. In this regard, we have also shown that the CD28/B7 costimulatory molecules play a critical role in TCRγδ cell activation [25].

Another consequence of antigen exposure to TCRγδ cells has been the observation that transgenic cells that bind antigen either in the thymus or epithelia, down-regulate Thy-1 expression [Tatsumi et al., unpublished observations]. This down-regulation is a direct con-

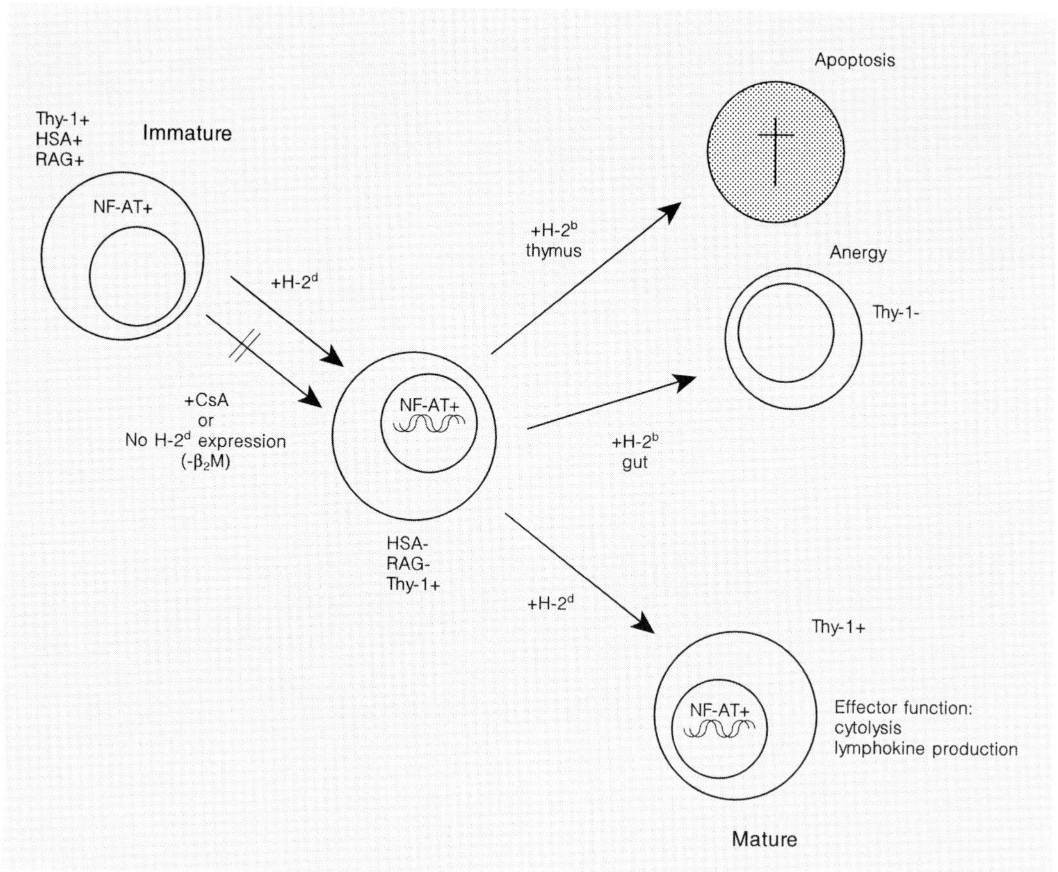

Fig. 1. TCRγδ selection. A schematic model of TCRγδ maturation from the immature HSA and RAG expressing thymocyte to the mature peripheral lymphocyte capable of full effector function is shown. In the absence of β_2M or in the presence of cyclosporine A (CsA), positive selection cannot occur and maturation is halted. This cell cannot proliferate to antigen; one reason may be the inability of NF-AT to bind its cognate DNA sequence and transactivate the expression of the IL-2 gene. In the presence of β_2M, the thymocytes mature. However, in the presence of high affinity antigen, negative selection occurs in the thymus by means of clonal deletion and in the gut epithelia by means of functional inactivation.

sequence of antigen exposure and correlates with T cell unresponsiveness. These results have general implications for the maintenance of self-tolerance of normal TCRγδ cells. Normal TCRγδ cells down-regulate Thy-1 after exposure to anti-TCR monoclonal antibodies in fetal thymus organ cultures. In fact, analysis of fetal development in the organ culture system has shown that the Vγ3 and Vγ4 subpopulation become Thy-1$^-$ during development. Thus, down-regulation of Thy-1 appears to be a marker of development and tolerance induction. Figure 1 is a schematic representation of the maturation pathway delineated from the study of the G8 transgenic mouse beginning from the immature (HSA^+, RAG^+, Thy-1^+) that is positively selected on H-2^d and develops into a mature effector cell (HSA^-, RAG^-, Thy-1^+) in the peripheral lymphoid organs capable to effector functions such as cytolysis and lymphokine production. These immature cells can be maintained when blocking positive selection with cyclosporine A or the down-regulation of H-2^d expression. Negative selection can be imposed when the ligand is of too high affinity, such as H-2^b in the case of G8 transgenic cells.

A second approach towards examining the development of TCRγδ cells has been to compare the repertoire of TCRγδ splenocytes in a variety of inbred and MHC-congenic strains of mice. A panel of antimurine TCRγδ mAb has allowed for more precise studies of the peripheral lymphoid organ TCRγδ repertoire in different strains of mice [26]. The percentages of both the $V\gamma2^+$ and $V\delta4^+$ subsets of splenic TCRγδ cells differ widely between different inbred strains of mice. Strikingly, the percentage of splenic Vγ2 cells in several strains of mice is quite high while in other strains of mice it is low (54% compared to 12%). Interestingly, the percentage of cells that express both of these V elements remains the same, suggesting that while the particular V genes are being selected for, this heterodimeric pair is not under selective pressure. Not surprisingly, these differences do not map to MHC genes or Mls determinants. F1 crosses between low and high expressing strains reveal that the Vγ2 high phenotype is dominant, suggesting a possible role for positive selection of these cells. In order to analyze whether these repertoire differences were due to the exogenous microbial insult, this analysis was also performed in germfree animals. Similar repertoire differences were observed, suggesting that the selection events were due to endogenous antigen(s) [Sperling et al., unpublished observations]. In addition, these selection events occur postnatally as the expansion of the respective subsets occurs between 1 and 4 weeks of age. While the origin of these respective subsets is dependent on the thymus, the actual expansion occurs in the periphery independently of the thymus, since expansion of the Vγ2 subset occurs in athymic mice. In order to map the genes responsible for the observed V gene segregation, recombinant inbred strains generated from a cross between high and low Vγ2 expressing mice were examined. By genetically classifying these strains, it became evident that high expression of Vγ2 mapped to the TCRδ locus while high expression of Vδ4 mapped to the TCRγ locus. This finding can be explained in at least two ways. First, pairing of V gene elements might have structural and biochemical restrictions that allow or prohibit particular heterodimers to be expressed. Second, the element that is being selected is the unknown V gene that is pairing with Vγ2 or Vδ4 that might be positively selected in the periphery. In this regard, there appears to be a limited junctional diversity in the expanded Vγ2 subset. V gene- and J gene-specific DNA oligos were used to amplify the junctions of the Vγ2 rearranged genes. Size differences in the CDR3 junctional region of these expressed gene elements were determined by resolving these amplified products on sequencing grade acrylamide gels. Interestingly, in the Vγ2 high expressing mouse strains, one major size is observed in neonates while a second population is apparent at week 1 of age (onset of peripheral expansion) and throughout adult life [Sperling et al., unpublished observations]. Furthermore, similar data is obtained at multiple time points in both thymic and splenic Vγ2 expressing cells. While these PCR products remain to be sequenced to confirm that a particular junctional sequence is being selected, it is evident that the selection event is restricting the size variation of this receptor and that there exists a newly expanded subset of Vγ2 expressing cells in the adult population. Finally, these results suggest that a positive selection event is occurring in the periphery of some strains of mice and is subsequently shaping the TCRγδ repertoire in the secondary lymphoid organs of these mice.

How Do TCRγδ Cells Recognize Nominal Antigen?

TCRγδ cells have been shown to recognize a variety of distinct antigens including MHC molecules, herpesvirus glycoprotein I, and heat shock proteins. However, the molecular nature of the antigenic ligands recognized by these TCRγδ cells has been unclear. Initial studies suggested that, unlike TCRαβ cells, TCRγδ cells may recognize their target antigens in a unique manner. First, the recognition of most antigens is MHC-unrestricted. In fact, even TCRγδ cells that react on MHC molecules cross-react on multiple haplotypes [27]. Second, few TCRγδ cells have been shown to directly recognize peptides in the context of MHC molecules. In those instances where peptides can be recognized, the precise involvement of the MHC has not been determined [28, 29]. Therefore, it is possible that TCRγδ cells do not recognize nominal antigens in a manner similar to TCRαβ cells. Studies in our laboratory have approached the role of peptide in the recognition of MHC-specific TCRγδ cells using two T cell clones. LBK5i (that recognizes a nonpolymorphic I-E epitope) and G8 (that recognizes a nonpolymorphic, nonclassical class I molecule, $T10^b$) [30]. These studies were initiated in collaboration with Dr. Yueh-hsiu Chien. In these studies, the G8 clone was shown to recognize the $T10^b$ glycoprotein expressed on transfected xenogeneic cells including human, hamster, simian monkey, and *Drosophila melanogaster*. Recognition of $T10^b$ on insect cells is significant since these cells are devoid of the classical peptide processing pathways and MHC molecules are transported to the cell surface devoid of peptides. Similarly, $T10^b$ can be presented to G8 when expressed on either RMA-S cells that have a characterized peptide transport deficiency that blocks peptide loading of MHC class I molecules or Chinese hamster ovary (CHO) cell ubiquitin mutants that cannot effectively process proteins into small peptides. Therefore, these data suggest that the G8 clone is recognizing an epitope of $T10^b$ that is expressed independently of peptide. This finding contrasts with another report that showed that mutations within the peptide binding groove of a related molecule, $T22^b$, abrogated recognition by another T-region-specific TCRγδ clone, KN6 [31]. However, it is uncertain whether the mutations in the $T22^b$ molecule affected the overall conformation of the molecule rather than simply dictating the peptide binding ability of the $T22^b$ molecule. In fact, recent data suggests that the $T10^b$ molecule is normally expressed on the cell surface devoid of peptides [Arunan, K., unpublished observations]. In collaboration with Dr. Donald Hunt at the University of Virginia, we have determined by mass spectroscopy that, in contrast to other class I molecules, the $T10^b$ molecules expressed on CHO cells do not bind small peptides. Therefore, it would appear that the TCRγδ cells are recognizing 'empty' $T10^b$. As importantly, it would appear that the $T10^b$ gene product does not associate with peptides on the cell surface, suggesting that these nonclassical MHC class I molecules may have a distinct function on the cell surface independent of peptidic antigen presentation.

Similar studies have been performed with the class II-specific T cell clone, LBK5i [30]. In those studies transfection of $I\text{-}E^k$ into the class II presentation mutant cell, T2, did not affect recognition by the clone. In addition, several point mutations within the peptide binding groove, that eliminated class II recognition by classical MHC-specific TCRαβ cells, did not alter I-E^k recognition by LBK5i. However, one point mutation outside the peptide binding groove abrogated recognition by LBK5i without affecting the recognition of I-E by cytochrome C-specific TCRαβ hybridomas. Finally, LBK5i was able to recognize purified soluble I-E^k devoid of peptides. These data parallel the data with the G8 clone in that the recognized epitopes are monomorphic and independent of peptides.

Thus, there is an emerging theory that the proteins recognized by TCRγδ cells can bypass classical antigen processing and presenting pathways and may be recognized as whole molecules in their native conformation. This notion is most apparent in a herpesvirus-specific T cell clone generated in our laboratory [32]. TgI4.4 was cloned from the draining lymph node of a C3H/HeJ mouse infected with herpesvirus type I and recognizes its cell surface glycoprotein, gI. This clone was found to recognize herpesvirus-infected target cells in an MHC class I and class II unrestricted manner. In fact, the clone recognized both class I- and class II-deficient cells. TgI4.4 reactivity is dependent, however, on cell surface expression of gI as truncation of gI that prevents membrane anchoring abrogates T cell recognition. Recognition was restored by anchoring the gI in the membrane with an irrelevant transmembrane amino acid sequence. In addition, gI-specific antiserum was able to block recognition. Taken together, these data suggested that TgI4.4 recognized gI in the absence of either a processing or presenting molecule. In fact, the herpesvirus-specific TCRγδ cells recognized a soluble form of the whole herpesvirus gI protein directly in an MHC- and antigen processing-independent way. A soluble form of gI was prepared by molecularly fusing the extracellular portion of gI onto an immunoglobulin Fc fragment to produce soluble gI-Ig. Immobilization of the gI-Ig fusion protein on plastic tissue culture wells stimulated the TCRγδ clone whereas a control fusion protein or soluble gI-Ig did not stimulate. This recognition is sensitive to reduction of the disulfide bonds and heat denaturation of gI. The recognition is also sensitive to trypsin digestion but is independent of N-linked glycosylation [Sciammas et al., unpublished observations]. One final example of nonconventional antigen recognition by TCRγδ cells is the recent identification of a small 2- to 10-kD protease-resistant phosphate containing molecule that is structurally analogous to TTP isolated from Mycobacteria. This compound was found to stimulate TCRγδ cells [33]. Antigen recognition was also MHC-unrestricted.

Together these results suggest that these TCRγδ cells exhibit recognition capabilities more similar to an immunoglobulin molecule where the antigen is recognized in a nonprocessed conformationally dependent form that is not complexed with a presenting molecule. In support of this idea, Rock et al. [4] reported a statistical analysis of CDR3 regions expressed by the seven antigen receptor chains (TCRα, β, γ, δ, IgH, Igκ, and Igλ). They showed that the distribution of CDR3 lengths of TCRδ and IgH are very diverse, ranging in length from very short to very long. In contrast, both the TCRα and β chains contain a size-restricted diversity in the CDR3 sequence while both TCRγ/δ chain pairs and IgH/L chain pairs can differ quite widely in their sizes. Thus, as seen in figure 2, the mechanism of antigen recognition utilized by TgI4.4, LBK5i and G8 may be representative of many TCRγδ cells. The recognition of whole glycoproteins may have important implications for the role of TCRγδ cells in disease inasmuch as the recognition of unprocessed antigens can occur rapidly and on anti-

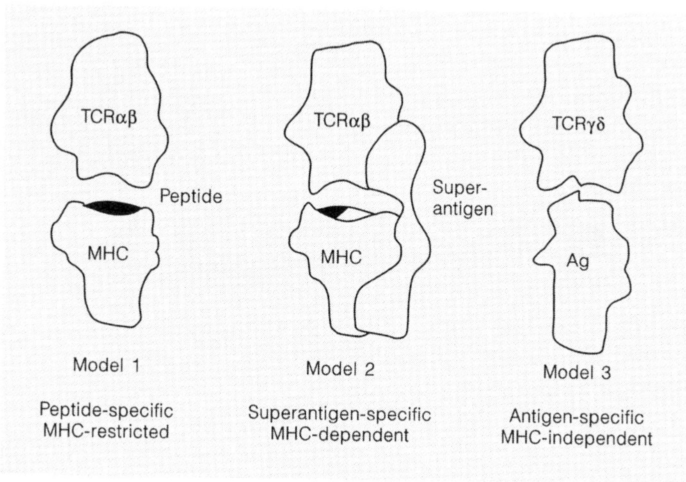

Fig. 2. Models of TCR recognition. Proposed here are three models by which TCRs interact with antigen (Ag). Model 1 describes the classical TCRαβ recognition of peptide restricted to MHC. Model 2 describes the interaction of the TCRβ chain with superantigen. While the recognition of superantigen is independent of antigen processing, it is dependent on MHC expression. Model 3 describes the TCRγδ interaction with unprocessed whole antigen in a manner much like immunoglobulin which is consistent with the clones examined to date.

gens expressed by cells that poorly express MHC molecules such as central nervous system cells or viral infected cells that prevent MHC transport to the cell surface.

What Role Do TCRγδ Cells Play in Disease or Regulation of Normal Immune Responses?

Several lines of evidence suggest that TCRγδ cells may play a critical role in both initiating and/or exacerbating local inflammatory responses. First, they are resident in the epithelia of the skin, lung and intestine, often at sites of initial disease insult. Second, TCRγδ cells are frequently observed at the site of inflammation and often expanded in infections such as HSV, HIV, influenza, *Listeria monocytogenes, Mycobacterium tuberculosis, Leishmania major, Shistosoma mansoni,* and *Trypanosoma cruzi* as well as in autoimmune diseases such as rheumatoid arthritis and celiac disease. Finally, TCRγδ cells produce multiple inflammatory lymphokines (e.g. IFNγ, TNF and TGFβ), express activation markers (CD25, CD44 and HSA), and in some instances, will lyse tumor cells when isolated directly out of the animal [1]. However, there has been little direct evidence for the role of TCRγδ cells in the pathogenesis of disease. The advent of both anti-TCRγδ monoclonal antibodies and the powerful approach of deleting individual TCR genes by homologous recombination for the analysis of the roles of TCRγδ cells in immune response has generated evidence that these cells both respond to and regulate pathogens.

Rheumatoid arthritis is an autoimmune condition in humans that is largely T cell-initiated. In mice, administration of purified type II collagen can initiate a disease (collagen-induced arthritis) that closely resembles the human condition. Since TCRγδ cells were identified in the inflamed synovial joints of arthritic patients, we have investigated their role in this mouse model [34]. Indeed, they are present, activated, and expanded in the joints of arthritic mice as well. In order to evaluate the role they play in arthritic pathology we used anti-TCRγδ antibodies to deplete TCRγδ cells prior to induction of disease [35].

These treated mice had a significantly delayed onset of disease and, in addition, the percentage of mice that had greater than two arthritic limbs was half the number observed in control mice. However, anti-TCRγδ treatment of mice during the course of disease greatly exacerbated the severity of the arthritis. This exacerbation was due to enhanced activation of the disease-activated TCRγδ cells in the synovial joints since mice treated with anti-TCRγδ $F(ab)'_2$ fragments developed disease normally. These data suggest that TCRγδ cells are involved in initiating the efficient progression to arthritis, and are active in the lymph nodes and synovial joints in mediating inflammation. Finally, the TCRγδ cells present in the inflamed joints expressed a limited receptor repertoire, expressing one of two TCRγ chains, Vγ1.1 and Vγ2. These results suggest that the TCRγδ cells participating in the inflammation are derived from the peripheral lymphoid organs and may be oligoclonal.

Listeria monocytogenes is a gram-positive rod that results in an intracellular infection of macrophages and hepatocytes. The murine model of listeriosis resembles human disease in terms of DTH responses, chronic progression of disease, and a dependence on T cells for clearance. The mechanism of clearance has been attributed to activation of macrophages by T cell-dependent factors and subsequent eradication of intracellular bacteria by oxidative cytotoxicity. We and others have observed an increase in the frequency of TCRγδ cells that migrate into the peritoneal cavity after infection [35, 36]. These activated TCRγδ cells secrete large amounts of IFNγ upon restimulation. They can be stimulated to proliferate to preparations of heat shock proteins as well as to components of heat-killed *Listeria* [36]. A direct demonstration of the role of TCRγδ cells in the immune response during the early stages of infection with *Listeria* was shown using antibody depletion and knockout technologies. Mice treated with anti-TCRγδ developed a much more severe infection as a consequence of an inability to suppress bacterial replication in infected spleens prior to the resolution of the disease by the TCRαβ cells. In contrast, depletion of the TCRαβ cells had no effect on the early replication of the bacteria but prevented the cure of the infection. Similar results were observed in TCRγδ and TCRαβ null mice [7]. Thus, it would appear that the TCRγδ cells enter the infected lesion rapidly, respond to exogenous or endogenous antigens including heat shock proteins and directly lyse the infected cells or produce inflammatory lymphokines such as IFNγ that activate resident macrophages to mediate the effector function.

Listeria is not the only infection that is regulated by TCRγδ cells. TCRγδ cells are prevalent in the parasitic infections induced by *Schistosoma mansoni, Leishmania major,* and *Mycobacterium tuberculosis.* The granulomas in the livers of *Schistosoma*-infected mice have a high frequency of TCRγδ cells [Sandor et al., personal commun.]. Moreover, the TCRs expressed by these granuloma-infiltrating cells express the characteristic TCR Vγ1.1 of the previously described hsp-reactive cells. Furthermore, we have shown that the migration of the TCRγδ cells to the granulomas is antigen-induced and independent of MHC class I expression. Finally, it appears as if *Schistosoma mansoni* induces both an early and a late wave of TCRγδ migration that are characterized by distinct TCRδ chain expression. Table 1 summarizes these data on TCRγδ cells and disease.

In conclusion, it is now clear that TCRγδ cells have an important function in the immune system. They express a limited, but specialized TCR repertoire, recognize antigens in a manner distinct from TCRαβ cells and act early in inflammatory lesions. However, several challenges and unanswered questions re-

Table 1. The phenotype and role of TCRγδ cells in collagen-induced arthritis, listeriosis, and schistosomiasis

	Activated phenotype	Receptor usage	Function	Antigenic specificity
Collagen-induced arthritis	$CD25^+$ $CD44^+$	Vγ1.1 Vγ2	increase severity	unknown CFA?
Listeriosis	$CD25^+$ $CD44^+$	Vγ1.1	reduce bacterial replication	HSP others?
Schistosomiasis	$CD25^+$ $CD44^+$	Vγ1.1 Vδ4 Vδ6	unknown	unknown

main in our pursuit of this area of research. Do the invariant TCR repertoires of the various epithelial subsets of TCRγδ suggest that these cells recognize an equally invariant array of antigens, with each subset recognizing a distinct antigen? Furthermore, the TCR repertoire of the peripheral TCRγδ cells suggests that they have the ability to recognize a wider array of antigens. What is the nature of these antigens? Are the antigens recognized as self or foreign? For instance, the large precursor population of hsp-reactive TCRγδ cells in the periphery can react to either stressed self tissues or to the highly evolutionarily conserved structures of hsps expressed by pathogens. Finally, the unique nature of antigen recognition, by at least some of these cells, suggests that they serve a unique role in immune responses. For example, TCRγδ cells could provide necessary cytokines in response to antigens that are not processed efficiently, antigens expressed in tissue that express MHC molecules poorly, or antigens that are potent in an unprocessed form.

References

1 Haas W, Pereira P, Tonegawa S: Gamma/delta T cells. Annu Rev Immunol 1993;11:637.
2 Allison JP, Havran WL: The immunobiology of T cells with invariant γδ antigen receptors. Annu Rev Immunol 1991;9:679.
3 Raulet DH: The structure, function, and molecular genetics of the gamma/delta T cell receptor. Annu Rev Immunol 1989;7:175.
4 Raulet DH: How γδ T cells make a living. Curr Biol 1994;4:246.
5 Born W, Happ MP, Dallas A, Reardon C, Kubo R, Shinnick T, Brennan P, O'Brien R: Recognition of heat shock proteins and γδ cell function. Immunol Today 1990;11:40.
6 Havran WL, Chien Y-H, Allison JP: Recognition of self antigens by skin-derived T cells with invariant γδ antigen receptors. Science 1991;252:1430.
7 Mombaerts P, Arnoldi J, Russ F, Tonegawa S, Kaufmann S: Different roles of αβ and γδ T cells in immunity against an intracellular bacterial pathogen. Nature 1993;365:53.
8 Tsuji M, Mombaerts P, Lefrancois L, Nussenzweig RS, Zavala F, Tonegawa S: γδ T cells contribute to immunity against the liver stages of malaria in αβ T-cell-deficient mice. Proc Natl Acad Sci USA 1994;91:345.
9 Rock EP, Sibbald PR, Davis MM, Chein Y-H: CDR3 length in antigen-specific immune receptors. J Exp Med 1994;179:323.

10 Houlden BA, Matis LA, Cron RQ, Widacki SM, Brown GD, Pampeno C, Meruelo D, Bluestone JA: A TCRγδ cell recognizing a novel TL-encoded gene product. Cold Spring Harbor Symp Quant Biol 1989;54:45.
11 Tonegawa S, Berns A, Bonneville M, Farr A, Ishida I, Ito K, Itohara S, Janeway CA Jr, Kanagawa O, Katsuki M, Kubo R, Lafaille J, Mombaerts P, Murphy D, Nakanishi N, Takagaki Y, Van Kaer L, Verbeek S: Diversity, development, ligands, and probable functions of γδ T cells. Cold Spring Harbor Symp Quant Biol 1989;54:31.
12 Yague J, White J, Coleclough C, Kappler JW, Palmer E, Marrack P: The T cell receptor: The alpha and beta chains define idiotype, and antigen and MHC specificity. Cell 1985;42:81.
13 Blackman M, Kappler J, Marrack P: The role of the T cell receptor in positive and negative selection of developing T cells. Science 1990;248:1335.
14 Asarnow DM, Cado D, Raulet DH: Selection is not required to produce invariant T cell-receptor gamma-gene junctional sequences. Nature 1993;362:158.
15 Iwashima MA, Green M, Bonyhadi M, Davis MM, Allison JP, Chien YH: Expression of a fetal γδ T-cell receptor in adult mice triggers a non-MHC-linked form of selective depletion. Int Immunol 1991;3:385.
16 Sim G-K, Augustin A: Dominant expression of the T cell receptor BALB invariant δ (BID) chain in resident pulmonary lymphocytes is due to selection. Eur J Immunol 1991;21:859.
17 Lefrancois L, LeCorre R, Mayo J, Bluestone JA, Goodman T: Extrathymic selection of TCR γδ T cells by class II major histocompatibility complex molecules. Cell 1990;63:333.
18 Parker CM, Groh V, Band H, Porcelli SA, Morita C, Fabbi M, Glass D, Strominger JL, Brenner MB: Evidence for extrathymic changes in the T cell receptor γδ repertoire. J Exp Med 1990;171:1597.
19 Wells FB, Gahm S-J, Hedrick SM, Bluestone JA, Dent AL, Matis LA: Requirements for positive selection of γδ receptor-bearing T cells. Science 1991;253:903.
20 Tatsumi Y, Pena JC, Matis L, Deluca D, Bluestone JA: Development of T cell receptor-γδ cells: Phenotypic and functional correlations of T cell receptor-gamma thymocyte maturation. J Immunol 1993;151:3030.
21 Jain J, McCaffrey PG, Miner Z, Kerppola TK, Lambert JN, Verdine GL, Curran T, Rao A: The T-cell transcription factor NF-ATp is a substrate for calcineurin and interacts with Fos and Jun. Nature 1993;365:352.
22 Dent AL, Matis LA, Hooshmand FH, Widacki SM, Bluestone JA, Hedrick SM: Self-reactive γδ T cells are eliminated in the thymus. Nature 1990;343:714.
23 Barrett TA, Delvy ML, Kennedy DM, Lefrancois L, Matis LA, Dent AL, Hedrick SM, Bluestone JA: Mechanism of self-tolerance of γδ T cells in epithelial tissue. J Exp Med 1992;175:65.
24 Sperling AI, Linsley PS, Barrett TA, Bluestone JA: CD28-mediated costimulation is necessary for the activation of T cell receptor-γδ$^+$ T lymphocytes. J Immunol 1993;151:6043.
25 Sperling AI, Cron RQ, Decker DC, Stern DA, Bluestone JA: Peripheral T cell receptor γδ variable gene repertoire maps to the T cell receptor loci and is influenced by positive selection. J Immunol 1992;149:3200.
26 Bluestone JA, Cron RQ, Barrett RA, Houlden B, Sperling AI, Dent A, Hedrick S, Fry A, Matis LA: Repertoire development and ligand specificity of murine TCRγδ cells. Immunol Rev 1991;120:5.
27 Schild H, Mavaddat N, Litzenberger C, Ehrich EW, Davis MM, Bluestone JA, Matis L, Draper RK, Chien Y: The nature of major histocompatibility complex recognition by γδ cells. Cell 1994;76:29.
28 Moriwaki S, Korn BS, Ichikawa Y, Van Kaer L, Tonegawa S: Amino acid substitutions in the floor of the putative antigen-binding site of H-2T22 affect recognition by a γδ T-cell receptor. Proc Natl Acad Sci USA 1993;90:11396.
29 Kozbor D, Trinchieri G, Monos DS, Isobe M, Russo G, Haney JA, Zmijewski C, Croce CM: Human TCRγδ$^+$, CD8$^+$ T lymphocytes recognize tetanus toxoid in an MHC-restricted fashion. J Exp Med 1993;169:1847.
30 Rivas A, Koide J, Cleary ML, Engleman EG: Evidence for involvement of the γδ T cell antigen receptor in cytotoxicity mediated by human alloantigen-specific T cell clones. J Immunol 1989;142:1840.
31 Sciammas R, Johnson RM, Sperling AI, Brady W, Linsley PS, Spear PG, Fitch FW, Bluestone JA: Unique antigen recognition by a herpesvirus-specific TCRγδ cell. J Immunol 1994;152:5392.
32 Constant P, Davodeau F, Peyrat MA, Poquet Y, Puzo G, Bonneville M, Fournie JJ: Stimulation of human γδ T cells by nonpeptide mycobacterial ligands. Science 1994;264:267.
33 Holoshitz J, Koning F, Coligan JE, De Bruyn J, Strober S: Isolation of CD4$^-$CD8$^-$ mycobacteria-reactive T lymphocyte clones from rheumatoid arthritis synovial fluid. Nature 1989;339:226.
34 Peterman GM, Spencer C, Sperling AI, Bluestone JA: Role of γδ T cells in murine collagen-induced arthritis. J Immunol 1993;151:6546.
35 Hiromatsu K, Yoshikai Y, Matsuzaki G, Ohga S, Muramori K, Matsumoto K, Bluestone JA, Nomoto K: A protective role of γδ T cells in primary infection with *Listeria monocytogenes* in mice. J Exp Med 1992;175:49.
36 Skeen MJ, Ziegler HK: Induction of murine peritoneal γδ T cells and their role in resistance to bacterial infection. J Exp Med 1993;178:971.

Immunol Res 1994;13:280–290

Harinder Singh

Department of Molecular Genetics and Cell Biology, Howard Hughes Medical Insitute, University of Chicago, Chicago, Ill., USA

Genetic Analysis of Transcription Factors Implicated in B Lymphocyte Development

Key Words
Transcription factors
Immunoglobulin genes
B lymphocyte development

Abstract
The transcription factors Oct-2, NF-κB and PU.1 have been implicated in regulating the development of B lymphocytes. Genetic approaches have been used to analyze the developmental functions of these regulatory proteins. Using gene targeting in murine embryonic stem cells, PU.1 is shown to be required for the development of progenitor B cells. Strikingly, PU.1 is also essential for the development of T lymphoid, granulocytic and monocytic progenitors. Transcription factors of the NF-κB/Rel family, which appear to regulate immunoglobulin kappa gene expression, are shown to be a target of the viral transforming protein (v-abl) which arrests B lineage development at the precursor B stage. This suggests a mechanism by which v-abl blocks precursor B cell differentiation. The Oct-2 transcription factor was considered to represent a development regulator of immunoglobulin gene expression. Using gene targeting in a murine B cell, Oct-2 is shown to be dispensable for immunoglobulin gene expression. This suggests the existence of an alternate pathway, involving the ubiquitous related protein, Oct-1, in immunoglobulin gene regulation.

The B cell developmental pathway represents a leading model for understanding the molecular basis of differentiation of a defined lineage in mammals [1]. B lymphocytes are produced from pluripotent hemopoietic stem cells in the fetal liver and the adult bone marrow. The developmental pathway is characterized by an ordered progression of DNA rearrangements that result in sequential expression of immunoglobulin (Ig) genes [2]. The heavy chain locus is preferentially activated for rearrangement in progenitor B cells.

Dr. Harinder Singh
Department of Molecular Genetics and Cell Biology
Howard Hughes Medical Institute
The University of Chicago, 920 East 58th Street
Chicago, IL 60637 (USA)

0257–277X/94/
0134–0280$8.00/0

Productive assembly of heavy chain gene segments leads to the generation of precursor B (pre-B) cells that express the heavy chain protein (μ) primarily in the cytoplasm. Pre-B cells express four additional lineage-restricted genes (VpreB, λ5, mb-1 and B29) that encode immunoglobulin (Ig)-associated proteins. The VpreB and λ5 products represent surrogate light chain proteins that associate with the μ protein to form a pre-B cell receptor, a small fraction of which is expressed on the cell surface. The mb-1 and B29 gene products are integral membrane proteins that are noncovalently associated with both the pre-B and the B cell Ig receptor complexes. The pre-B cell receptor complex has been genetically shown to be required for further development [3]. It therefore represents an important developmental checkpoint in the pathway. Pre-B cells that fail to perform productive heavy chain gene rearrangement are eliminated by programmed cell death. On the other hand, intermediates which productively assemble heavy chain gene segments appear to be selected for survival and further differentiation through the expression of the pre-B receptor. Pre-B cells require stromal cell contact and interleukin-7 (IL-7) for their development and proliferation. To differentiate into antigen receptor-expressing B lymphocytes, pre-B cells must activate rearrangement of one of two light chain loci (κ or λ). The vast majority of murine B cells (95%) rearrange and express the kappa locus. B cells are triggered by antigen binding and T cell signals to terminally differentiate into Ig-secreting cells.

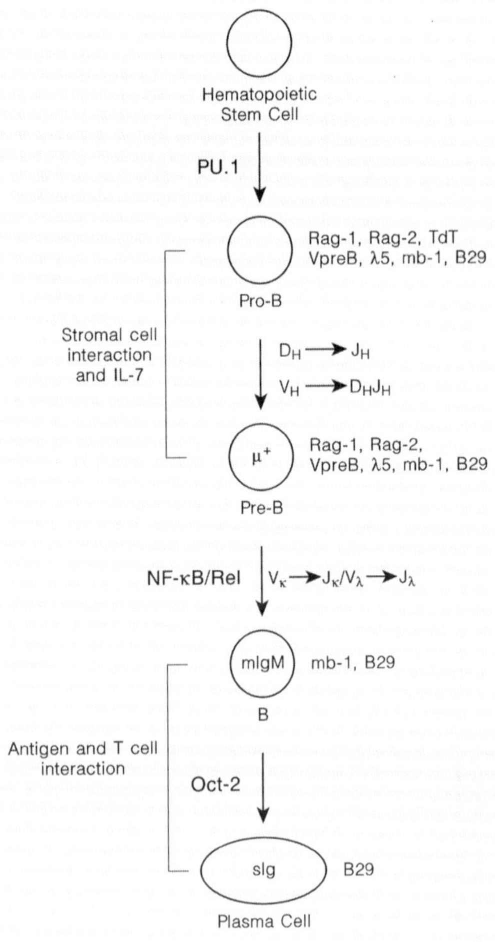

Fig. 1. The B lineage developmental pathway.

Transcription factors represent key components of regulatory networks underlying developmental pathways. My laboratory has explored the biological functions and regulation of three distinct transcription factors (Oct-2, NF-κB/Rel and PU.1) that have been implicated in the development of B lymphocytes. Figure 1 depicts the B lineage developmental pathway and highlights the stages at which the aforementioned transcription factors appear to execute critical regulatory functions. Using gene targeting in mouse embryonic stem cells, the laboratory has recently shown that the PU.1 transcription factor is required for the development of progenitor B cells [4]. Strikingly, PU.1 is also shown to be required for the development of T lymphoid, granulocytic and monocytic progenitors. This study represents the first demonstration of a common genetic requirement for the development of lymphoid and myeloid lineages. Transcription factors of the NF-κB/Rel family have been suggested to regulate the differentiation of pre-B cells by controlling the developmental activation of the kappa light chain locus [5]. Our work has strongly strengthened this proposal by providing the first demonstration of activated NF-κB/Rel complexes correlating with kappa transcription in two types of nontransformed pre-B cells. Most importantly, this study has discovered an unsuspected inhibitory interaction between the viral transforming protein (v-abl), which arrests B lineage development at the pre-B stage, and the NF-κB/Rel family of transcription factors [6]. This analysis has suggested a molecular pathway by which v-abl inhibits kappa locus transcription and rearrangement. The Oct-2 transcription factor was considered to represent a developmental regulator of Ig gene expression by virtue of its interaction with a functionally essential octamer element found in Ig gene promoters. We have genetically analyzed Oct-2 function by using gene targeting to disrupt both alleles of the locus in a murine B cell. In spite of a dramatic reduction in Oct-2 protein levels, no alteration in Ig gene expression was observed. This result represents the first genetic demonstration that Oct-2 is not necessary for the maintenance of Ig gene expression in a differentiated B cell [7]. It suggests the existence of an alternate pathway, involving the ubiquitous related protein, Oct-1, in Ig gene regulation. The conclusion that an Oct-2-independent pathway can promote Ig gene transcription in B cells has been corroborated by targeting of the Oct-2 gene in embryonic stem cells and the generation of homozygous Oct-2 null mice [8].

This study by the Baltimore laboratory showed that the Oct-2 gene is essential for viability (the mutant mice die within a day of birth), but normal numbers of surface Ig-positive B cells develop in the fetal liver. However, differentiation, in vitro, of Ig-positive B cells into Ig-secreting plasma cells is severely inhibited. Thus, a deficiency in the Oct-2 regulatory protein appears to result in the generation of nonfunctional B lymphocytes. These two genetic studies have provided a new perspective on the role of Oct-2 in B cell development. They have shown that Oct-2 is not essential for regulating the expression of Ig genes and therefore does not act at early stages of the developmental pathway. Nevertheless, Oct-2 is a nonredundant activator in B lineage cells and is required for terminal differentiation of B lymphocytes. These research contributions by my laboratory are expanded below.

Genetic Analysis of Oct-2 Reveals Two Distinct, Cell Type-Specific Pathways of Octamer-Dependent Gene Activation in B Cells

Ig gene promoters (μ, κ and λ) are preferentially active in B lineage cells [9–12]. They function in concert with Ig gene enhancers to direct B lineage-restricted transcription of Ig genes [13]. A highly conserved octamer element, ATTTGCAT, positioned ~ 70 bp upstream of the transcriptional start site, is the major regulatory element of Ig gene promoters [14]. It is essential for activity of Ig gene promoter constructs transfected into B cells [11, 15, 16]. Mutation of this site in the promoter of a rearranged Ig heavy chain gene in transgenic mice reduces expression 20- to 30-fold in lymphoid tissues [17]. Furthermore, the octamer element confers B cell-specific activity on minimal heterologous promoters when placed ~ 70 bp upstream from the transcriptional start site [18, 19]. Thus, the octamer element appears to be necessary and sufficient to account for the B cell-specific activity of Ig promoters. Paradoxically, the same motif is an important element of the promoters of ubiquitously expressed genes such as the U2 and U6 snRNA genes and the histone H2B gene [20, 21]. In these promoter contexts, the octamer element stimulates transcription in various non-B lineage cells.

The ability of the octamer element to promote ubiquitous as well as B cell-specific gene expression was initially suggested to reside in its interaction with two distinct transcription factors Oct-1 and Oct-2. The two factors have indistinguishable DNA-binding specificities, but are products of distinct genes [22–27]. Oct-1 and Oct-2 are members of the POU family of homeodomain proteins [28]. POU proteins contain a 150- to 160-amino acid bipartite DNA-binding domain which consists of a conserved N-terminal POU-specific segment and a carboxy-terminal homeodomain [29]. Oct-1 and Oct-2 represent a prototypical example of distinct transcription factors which function through interaction with a common motif. The Oct-1 gene is expressed in a variety of cell types, whereas Oct-2 expression is restricted to cells of the lymphoid and central nervous systems [25, 27, 30]. Their differing distributions led to the proposal that Oct-1 regulates transcription of U snRNA and histone H2B genes, while Oct-2 specifically activates Ig gene transcription [31]. The proposal that Oct-2 specifically promotes Ig gene transcription has been most strongly challenged by the following observations of several in vitro transcription analyses [32–34]: (i) although Ig gene promoters are preferentially active in nuclear extracts derived from B lineage cells, high level promoter activity does not correlate with corresponding Oct-2 levels; (ii) highly purified Oct-1 and Oct-2 are equally efficient at stimulating transcription from Ig gene promoters in nuclear extracts from non-B or B cells depleted of octamer binding proteins, and (iii) high level Ig gene promoter activity in B cell nuclear extracts appears to require an additional B-cell factor which can interact with Oct-1 or Oct-2. These observations have led to the suggestion that Ig gene promoter activity may be regulated by a pathway involving Oct-1 or Oct-2 and a common B cell accessory factor [34]. Recently the factor responsible for this activity (OCA-B) has been partially purified and characterized [35].

To genetically test the function(s) of Oct-2 we have disrupted both copies of the wild type Oct-2 allele by gene targeting in a murine B cell. In spite of a drastic reduction in Oct-2 levels (20-fold) no effect was observed on the expression of endogenous Ig genes. Expression of transiently transfected reporter genes containing either an Ig gene promoter or a heterologous promoter with a single octamer element was also unaffected by the reduction in Oct-2 levels. In contrast to the above, expression of a reporter construct containing multiple octamer motifs upstream of a heterologous promoter was severely reduced (10-fold) in the double disruptant cells. These results provided the first genetic support for the existence of an Oct-2-independent B cell-specific pathway for Ig gene transcription, which likely involves Oct-1 and the B cell coactivator OCA-B [7]. They also genetically revealed a distinct Oct-2-dependent pathway of octamer-mediated gene activation. Our conclusion that an Oct-2-independent pathway can promote Ig gene transcription in B cells was corroborated by targeting of the Oct-2 gene in embryonic stem cells and the generation of homozygous Oct-2 null mice [8]. The Oct-2 gene is essential for viability (the mutant mice die within a day of birth), but normal numbers of surface Ig-positive B cells develop in the fetal liver. Ig gene (μ as well as κ) transcription in fetal liver-derived pre-B cell lines occurs at wild type levels. However, differentiation, in vitro, of Ig-positive B cells into Ig-secreting plasma cells is severely inhibited. Thus Oct-2 is critically required for the terminal differentiation of B lymphocytes.

Our Oct-2-targeted B cells exhibit a molecular defect. They poorly activate transcription of a reporter construct containing a multiple array of octamer elements upstream of a heterologous promoter. Thus Oct-2 is not a redundant activator protein in a B cell. This suggests that there may exist B cell-specific genes which critically depend on Oct-2 for their activity. The transcriptional regulatory regions of these genes may contain a multiple array of octamer elements either upstream or downstream from their promoters. To our knowledge, this study represents the first example of targeting both alleles of a diploid locus in a somatic mammalian cell line. The same approach can be extended to other genes thereby greatly enhancing the power of somatic cell genetics. Extension of this approach to genes encoding other transcription factors will allow a genetic dissection of their functions within the context of cell lines representing various differentiation states. Such an approach will complement the developmental analysis of these regulatory proteins via gene targeting in embryonic stem cells particularly in cases of embryonic lethality or defects manifested early in cellular differentiation.

Regulation of Oct-2 Transcription in B Lineage Cells

The Oct-2 gene is expressed in a developmentally regulated manner in the lymphoid and central nervous systems [25, 30]. In B lineage cells, my laboratory has shown that expression of the Oct-2 gene is transcriptionally upregulated during pre-B cell differentiation [36]. Importantly, increased expression of the gene occurs just before the stage in development when the function of Oct-2 is required. This suggests that upregulation of Oct-2 transcription may be an obligatory event for B cell development. By analyzing transcription factors that regulate Oct-2 expression in B-lineage cells we hope to identify other regulatory genes that constitute members of the genetic hierarchy controlling B cell development. The effort to identify the Oct-2 gene promoter was made difficult by the fact that the gene has a very complex pattern of transcripts with at least five distinct mRNAs, ranging in size from 2 to 8 kb. The targeting of the first coding exon of the Oct-2 gene in my laboratory unequivocally established that the locus has a single major promoter. Utilizing this information the laboratory has cloned the Oct-2 promoter region. The promoter has been characterized using functional assays. Although the promoter is preferentially active in B cells, significant activity is also detectable in non-B lineage cells. This strongly suggests an additional level of negative regulation of Oct-2 transcription since expression of the gene is tightly regulated in a cell type-specific manner. Most importantly, the characterization of the promoter has led to discovery of a lipopolysaccharide (LPS)-inducible transcription factor that binds to the Oct-2 promoter. This factor likely plays a critical role in upregulating Oct-2 expression in differentiating pre-B cells. Thus, it may represent an upstream member of the regulatory protein network underlying B cell development.

The v-abl Tyrosine Kinase Negatively Regulates NF-κB/Rel Factors and Blocks kappa Gene Transcription in Pre-B Lymphocytes

Transcriptional activation and rearrangement of the kappa light chain locus represent pivotal regulated events in the B cell developmental pathway. Transcription of the kappa locus is controlled by two enhancers, the intron enhancer ($E_{i\kappa}$) and the 3′ enhancer ($E_{3'\kappa}$), which flank the constant region exon [13]. The intron enhancer appears to regulate the developmental activation of the kappa locus; it functions as a strongly inducible enhancer in transformed pre-B cell lines in which transcription of germ line kappa alleles is activated by bacterial LPS stimulation [37, 38]. $E_{i\kappa}$ also appears to regulate rearrangement of the kappa locus. Targeted deletion of $E_{i\kappa}$ in murine embryonic stem cells results in a complete block to rearrangement of the kappa locus in B lineage descendants of the mutant stem cells [39].

Transcription factors of the NF-κB/Rel family regulate the activity of the kappa intron enhancer [40, 41]. They bind to the κB site in $E_{i\kappa}$, which is essential for the LPS-inducible activity of the enhancer in transformed pre-B cell lines [37, 38]. NF-κB is a heterodimer comprised of two subunits, p50 and p65. The κB site is additionally recognized by a diverse set of heterodimeric or homodimeric complexes including c-rel, rel B and p52 subunits. In unstimulated pre-B cell lines, NF-κB/Rel complexes are sequestered in the cytoplasm in an inactive state by inhibitory IκB subunits [42, 43]. LPS stimulation results in highly accelerated degradation of IκBα, thereby allowing nuclear translocation of NF-κB/Rel complexes and their binding to the kappa intron enhancer [44, 45]. NF-κB/Rel complexes are inducible in a wide variety of cell types by a diverse set of signals including mitogens and inflammatory cytokines. However, the nature of the developmental signaling pathway that activates NF-κB/Rel complexes in nontransformed pre-B cells is obscure. Pre B cell lines generated by in vitro transformation of fetal liver or adult bone marrow cells with the Abelson murine leukemia virus have been extensively used as models to analyze mechanisms regulating kappa gene transcription and rearrangement [46]. In almost every case, these cell lines have initiated or completed rearrangements at a transcriptionally active heavy chain locus but contain germline kappa alleles that are transcriptionally inactive [47, 48]. A longstanding question has been if transformation by the virus actively blocks transcription and/or rearrangement of the kappa locus. We have gained insight into the mode of activation of NF-κB/Rel factors in differentiating pre-B cells. Additionally we have discovered an unanticipated inhibitory interaction between the viral transforming protein v-abl and the NF-κB/Rel factors which could account for the transformation-induced developmental block [6].

This study was initiated by our demonstration that nontransformed pre-B cells expanded from the mouse bone marrow efficiently transcribe unrearranged kappa alleles. In addition, they contain activated complexes of the NF-κB/Rel transcription factor family, in contrast with their Abelson-transformed counterparts. These observations led us to consider the hypothesis that transformation of pre-B cells by v-abl might result in a block of germ line kappa gene transcription as a consequence of negative regulation of NF-κB/Rel activity. Using conditionally transformed pre-B cell lines, we have shown that v-abl, a tyrosine kinase, blocks germ line kappa gene transcription and negatively regulates NF-κB/Rel activity. An active v-abl kinase specifically inhibits the NF-κB/Rel-dependent kappa intron enhancer, which is implicated in promoting both transcription as well as rearrangement of the kappa locus. v-abl inhibits the activated state of NF-κB/Rel complexes in a pre-B cell via a post-translational mechanism that results in increased stability of the inhibitory subunit IκBα. This analysis suggests a molecular pathway by which v-abl inhibits kappa locus transcription and rearrangement.

In this study, we have demonstrated activated NF-κB/Rel complexes correlating with germ line kappa transcription in two types of nontransformed pre-B cells. The first type is represented by pre-B cells expanded from the mouse bone marrow in the presence of stromal cells and IL-7. The second type is represented by conditionally transformed pre-B cells that have been reverted to their nontransformed state. Importantly, activation of NF-κB/Rel factors in the latter cells is observed in the absence of stromal cells and IL-7. This suggests that the activation of NF-κB/Rel factors in pre-B cells is an intrinsic feature of their differentiation state. Therefore, during B cell development, NF-κB/Rel factors may be activated as a secondary response (dependent upon protein synthesis) to an extracellular differentiation signal derived from the stroma. For example, stromal cell-dependent development of pre-B cells could involve increased expression of an IκBα kinase gene. This would promote increased dissociation and turnover of IκBα, thereby activating NF-κB/Rel factors. Our results suggest that the NF-κB/Rel system may have evolved two distinct modes of regulation for executing different functions during cellular activation and differentiation. In the former situation, NF-κB/Rel factors are activated as a primary response (independent of protein synthesis) to an extracellular signal such as tumor necrosis factor α. Signaling results in highly accelerated turnover of IκBα, which promotes a rapid and short-lived activated state. In the latter situation, NF-κB/Rel factors are activated as a secondary response to a differentiation signal. Differentiation results in a modest increase in IκBα turnover. This promotes the generation of a slower but extended activated state.

Our results with bone marrow-derived pre-B cell cultures provide evidence for a normal stage of B cell development in which activation of germ line kappa transcription is uncoupled from kappa light chain rearrangement. This strongly suggests that the two events are independently regulated and therefore activated by distinct developmental signals. The IL-7-expanded pre-B cells appear to represent a novel intermediate in development since kappa transcription but not rearrangement has been activated. It will be of significant interest to determine what signal(s) can induce kappa rearrangement in these normal pre-B cells particularly in light of the observation that proficient rearrangement is activated simply by inactivating v-abl in conditionally transformed pre-B cells [49]. Both the IL-7-expanded and the conditionally transformed v-abl^{ts} pre-B cells represent new systems to explore the signaling pathway(s) and mechanism(s) by which kappa rearrangement is induced during B lymphocyte development.

The Transcription Factor PU.1/Spi-1 Is Required for the Development of Multiple Hematopoietic Lineages

The transcription factor PU.1 is the product of the *Spi-1* proto-oncogene [50, 51]. The *Spi-1* locus is the site of integration of the spleen focus-forming proviurs in 95% of murine erythroleukemias induced by Friend virus complexes [52]. Spleen focus-forming proviral insertion leads to overexpression of PU.1 in erythroblasts which has been shown to be sufficient for their immortalization [53]. Binding sites for PU.1 are present in intron 2 of the mouse β-major (β^M)-globin gene [54]. This region may regulate β-globin gene expression since it has an altered chromatin structure in erythroid cells. Therefore, PU.1 may be required for the development of erythroid progenitors during hematopoiesis.

PU.1 is a member of the *ets* family of transcription factors [55]. It was independently cloned on the basis of its binding to a purine-rich sequence, GAGGAA (PU box), in a MHC class II gene promoter [50, 51]. The DNA binding domain (*ets* homology region) of PU.1 is located in the highly basic carboxy-terminal region. The *PU.1* gene is expressed specifically in hematopoietic tissues, with particularly high levels of expression in the monocytic and B-lymphoid lineages [50, 51, 56]. Numerous presumptive PU.1 target genes have been identified in the B and monocytic lineages. In the B lineage PU.1 is implicated in regulating transcription of the Ig heavy (μ) [57] and light chain genes (κ and λ) [58, 59], the *mb-1* gene [60, 61], and the J-chain gene [62]. Functionally important PU.1 binding sites are present in the enhancers or promoters of these genes. PU.1 recruits a distinct transcription factor, NF-EM5, to the Ig kappa 3′ and lambda 2–4 enhancers [58, 59]. In monocytes, PU.1 appears to regulate transcription of the genes encoding the macrophage colony-stimulating factor receptor and the myeloid integrin CD11b [63, 64]. Based on these studies, PU.1 has been suggested to control the differentiation of B lymphocytes and monocytes.

To genetically analyze the function(s) of PU.1, we have used gene targeting in embryonic stem cells to generate mice carrying a mutant allele. *PU.1* homozygous mutant embryos die at a late gestational stage. Mutant embryos produce normal numbers of megakaryocytes and erythroid progenitors, but some show an impairment of erythroblast maturation. A most striking and unexpected consequence of the mutation is an invariant multilineage defect in the generation of B and T lymphocytic, monocytic, and granulocytic progenitors [4]. For each lineage, analysis with multiple stage-specific markers failed to reveal any differentiating intermediates. Since erythroid progenitors are generated in normal numbers in the *PU.1* mutant embryos, this multilineage defect is not simply a reflection of a *PU.1* requirement for the functioning of the hematopoietic stem cell. Although effects on B cell and monocyte development could be anticipated from previous work the multilineage phenotype is novel at two levels: (i) it shows a requirement for PU.1 for the generation of T lineage and granulocytic progenitors and (ii) it reveals that in the B lineage, PU.1 is not simply required for the activation of Ig heavy and kappa light chain gene transcription. Instead, PU.1 is required for the development of B lineage progenitors that undergo Ig gene rearrangements and express the cell surface markers B220 and CD43 and the genes *mb-1, B29,* and *VpreB*. This latter result is particularly noteworthy since it reveals that PU.1 is a pleiotropic regulator of B cell development acting at the earliest identifiable stage. In this regard, it is important to note that Ig gene rearrangement events can be genetically uncoupled from the expression of the cell surface markers B220 and CD43 and the genes *mb-1, B29,* and *VpreB* in B lineage progenitors by the *Rag-1* and *Rag-2* mutations [65, 66].

To date only a few genes, *c-myb, Rag-1* and *Rag-2*, have been rigorously shown to be required for the development of multiple, but not all, lineages of the hematopoietic system. Mutation of the *c-myb* locus blocks the development of erythroid and myeloid lineages but not of megakaryocytes [67]. *Rag-1* and *Rag-2* are required for the development of B and T lymphocytes [65, 66]. The phenotype of the *PU.1* mutation is unique among this select group of genes. It, for the first time, establishes a common genetic requirement for the development of lymphoid and myeloid lineages. The simplest interpretation of our results would suggest the existence of a multipotent *PU.1*-dependent lymphoid-myeloid progenitor. Bone marrow-derived clones representing such an intermediate in the hematopoietic system have recently been described [68]. Furthermore, in vitro clonogenic assays have revealed progenitors with both myeloid and B-lymphoid potential [69, 70].

Acknowledgments

This work was supported by the Howard Hughes Medical Institute. I would like to thank Gail Hubbard for preparation of the manuscript.

References

1 Rolink A, Melchers F: Molecular and cellular origins of B lymphocyte diversity. Cell 1991;66:1081–1094.

2 Blackwell TK, Alt FW: Mechanism and developmental program of immunoglobulin gene rearrangement in mammals. Annu Rev Genet 1989;23:605–636.

3 Ehlich A, Schaal S, Gu H, Kitamura D, Müller W, Rajewsky K: Immunoglobulin heavy and light chain genes rearrange independently at early stages of B cell development. Cell 1993;72:695–704.

4 Scott EW, Simon MC, Anastasi J, Singh H: Requirement of transcription factor PU. 1 in the development of multiple hematopoietic lineages. Science 1994;265:1573–1577.

5 Liou H, Baltimore D: Regulation of the NF-κB/rel transcription factor and IκB inhibitor system. Curr Opin Cell Biol 1993;5:4777–487.

6 Klug CA, Gerety SJ, Shah P, Chen Y, Rice NR, Rosenberg N, Singh H: The *v-abl* tyrosine kinase negatively regulates NF-κB/Rel factors and blocks κ gene transcription in pre-B lymphocytes. Genes Dev 1994;8:678–687.

7 Feldhaus AL, Klug CA, Arvin KL, Singh H: Targeted disruption of the Oct-2 locus in a B cell provides genetic evidence for two distinct cell type-specific pathways of octamer element-mediated gene activation. EMBO J 1993;12:2763–2772.

8 Corcoran LM, Karvelas M, Nossal GJV, Ye Z-S, Jacks T, Baltimore D: Oct-2, although not required for early B-cell development, is critical for later B-cell maturation and for postnatal survival. Genes Dev 1993;7:570–582.

9 Foster J, Stafford J, Queen C: An immunoglobulin promoter displays cell-type specificity independently of the enhancer. Nature 1985;315:423–425.

10 Gopal TV, Shimada T, Bauer AW, Nienhuis AW: Contribution of promoter to tissue-specific expression of the mouse immunoglobulin kappa gene. Science 1985;229:1102–1105.

11 Mason JO, Williams GT, Neuberger MS: Transcription cell type specificity conferred by an immunoglobulin V_H gene promoter that includes a functional consensus sequence. Cell 1985;41:479–487.

12 Picard D, Schaffner W: Cell-type preference of immunoglobulin κ and λ gene promoters. EMBO J 1985;4:2831–2838.

13 Staudt LM, Lenardo MJ: Immunoglobulin gene transcription. Annu Rev Immunol 1991;9:373–398.

14 Parslow TG, Blair DL, Murphy WJ, Granner DK: Structure of the 5′ ends of immunoglobulin genes: A novel conserved sequence. Proc Natl Acad Sci USA 1984;81:2650–2654.

15 Bergman Y, Rice D, Grosschedl R, Baltimore D: Two regulatory elements for immunoglobulin κ light chain gene expression. Proc Natl Acad Sci USA 1984;81:7041–7045.

16 Falkner FG, Zachau HZ: Correct transcription of an immunoglobulin κ gene requires an upstream fragment containing conserved sequence elements. Nature 1984;310:71–74.

17 Jenuwain T, Grossehedl R: Complex pattern of immunoglobulin mu gene expression in normal and transgenic mice: Nonoverlapping regulatory sequences govern distinct tissue specificities. Genes Dev 1991;5:932–943.

18 Dreyfus M, Doyen N, Rougeon F: The conserved decanucleotide from the immunoglobulin heavy chain promoter induces a very high transcriptional activity in B-cells when introduced into a heterologous promoter. EMBO 1987;6:1685–1690.

19 Wirth T, Staudt L, Baltimore D: An octamer oligonulceotide upstream of a TATA motif is sufficient for lymphoid-specific promoter activity. Nature 1987;329:174–178.

20 LaBella F, Sive HL, Roeder RG, Heintz N: Cell-cycle regulation of a human histone H2b gene is mediated by the H2b subtype-specific consensus elements. Genes Dev 1988;2:32–39.

21 Murphy S, Moorefield B, Pieler T: Common mechanisms of promoter recognition by RNA polymerases II and III. Trends Genet 1989;5:122–126.

22 Singh H, Sen R, Baltimore D, Sharp PA: A nuclear factor that binds to a conserved sequence motif in transcriptional control elements of immunoglobulin genes. Nature 1986;319:154–158.

23 Staudt LM, Singh H, Sen R, Wirth T, Sharp PA, Baltimore D: A lymphoid-specific protein binding to the octamer motif of immunoglobulin genes. Nature 1986;323:640–643.

24 Müller MM, Ruppert S, Schaffner W, Matthias P: A cloned octamer transcription factor stimulates transcription from lymphoid-specific promoters in non-B cells. Nature 1988;336:544–551.
25 Staudt LM, Clerc RG, Singh H, LeBowitz JH, Sharp PA, Baltimore D: Cloning of a lymphoid specific cDNA encoding a protein binding the regulatory octamer motif. Science 1988;241:577–580.
26 Scheidereit C, Cromlish JA, Gerster T, Kawakami K, Balmaceda C, Currie RA, Roder RG: A human lymphoid-specific transcription factor that activates immunoglobulin genes is a homeobox protein. Nature 1988;336:551–557.
27 Sturm RA, Das G, Herr W: The ubiquitous octamer protein Oct-1 contains a POU domain with a homeo subdomain. Genes Dev 1988;2:1582–1599.
28 Herr W, Sturm RA, Clerc RG, Corcoran LM, Baltimore D, Sharp PA, Ingraham HA, Rosenfeld MG, Finney M, Ruvkun G, Horvitz RH: The POU domain: A large conserved region in the mammalian pit-1, oct-1, oct-2, and Caenorhabditis elegans *unc*-86 gene products. Genes Dev 1988;2:1513–1516.
29 Verrijzer CP, Alkema MJ, van Weperen WW, Van Leeuwen HC, Strating MJJ, van der Vliet PC: The DNA binding specificity of the bipartite POU domain and its subdomains. EMBO J 1992;11:4993–5003.
30 He X, Treacy N, Simmons DM, Ingraham HA, Swanson LW, Rosenfeld MG: Expression of a large family of POU-domain regulatory genes in mammalian brain development. Nature 1989;340:35–42.
31 Schaffner W: How do different transcription factors binding the same DNA sequence sort out their jobs? Trends Genet 1989;5:37–39.
32 LeBowitz JH, Kobayashi T, Staudt L, Baltimore D, Sharp PA: Octamer-binding proteins from B of HeLa cells stimulate transcription of the immunoglobulin heavy-chain promoter in vitro. Genes Dev 1988;2:1227–1237.
33 Johnson DG, Carayannopoulos L, Capra JD, Tucker PW, Hanke JH: The ubiquitous octamer-binding protein(s) is sufficient for transcription of immunoglobulin genes. Mol Cell Biol 1990;10:982–990.
34 Pierani A, Heguy A, Fujii H, Roeder RG: Activation of octamer-containing promoters by either octamer-binding transcription factor 1 (OTF-1) or OTF-2 and requirement of an additional B-cell-specific component for optimal transcription of immunoglobulin promoters. Mol Cell Biol 1990;10:6204–6215.
35 Luo Y, Fujii H, Gerster T, Roeder RG: A novel B cell-derived coactivator potentiates the activation of immunoglobulin promoters by octamer-binding transcription factors. Cell 1992;71:231–241.
36 Miller CL, Feldhaus A, Rooney JW, Rhodes LD, Sibley CH, Singh H: Regulation and a possbile stage-specific function of Oct-2 during Pre-B-cell differentiation. Mol Cell Biol 1991;11:4885–4894.
37 Atchison ML, Perry RP: The role of the κ enhancer and its binding factor NF-κB in the developmental regulation of κ gene transcription. Cell 1987;48:121–128.
38 Lenardo M, Pierce JW, Baltimore D: Protein-binding sites in Ig gene enhancers determine transcriptional activity and inducibility. Science 1987;236:1573–1577.
39 Takeda S, Zou Y-R, Bluethmann H, Kitamury D, Muller U, Rajewsky K: Deletion of the immunoglobulin κ chain intron enhancer abolishes κ chain gene rearrangement in *cis* but not λ chain gene rearrangement *in trans*. EMBO J 1993;12:2329–2336.
40 Lenardo M, Baltimore D: NF-κB: A pleiotropic mediator of inducible and tissue-specific gene control. Cell 1989;58:227–229.
41 Blank V, Kourilsky P, Israël A: NF-κB and related proteins: Rel/dorsal homologies meet ankyrin-like repeats. TIBS 1992;17:135–140.
42 Baeuerle PA, Baltimore D: IκB: A specific inhibitor of the NF-κB transcription factor. Science 1988;242:540–545.
43 Beg AA, Baldwin AS: The IκB proteins: Multifunctional regulators of Rel/NF-κB transcription factors. Genes Dev 1993;7:2064–2070.
44 Henkel T, Machleidt T, Alkalay I, Kronke M, Ben-Neriah Y, Baeuerle PA: Rapid proteolysis of IκB-a is necessary for activation of transcription factor NF-κB. Nature 1993;365:182–185.
45 Rice NR, Ernst MK: In vivo control of NF-κB activation by IκBα. EMBO 1993;12:4685–4695.
46 Baltimore D, Rosenberg N, Witte, ON: Transformation of immature lymphoid cells by Abelson murine leukemia virus. Immunol Rev 1979;48:3–22.
47 Alt F, Rosenberg N, Lewis S, Thomas E, Baltimore D: Organization and reorganization of immunoglobulin genes in A-MuLV-transformed cells: Rearrangement of heavy but not light chain genes. Cell 1981;27:381–390.
48 Nelson KJ, Kelley DE, Perry RP: Inducible transcription of the unrearranged κ constant region locus is a common feature of pre-B cells and does not require DNA or protein synthesis. Proc Natl Acad Sci USA 1985;82:5305–5309.
49 Chen Y-Y, Wang LC, Huang M-S, Rosenberg N: An active v-abl protein tyrosine kinase blocks immunoglobulin light-chain gene rearrangement. Genes Dev 1994;8:688–697.
50 Klemsz MJ, McKercher SJ, Celada A, Van Beveren C, Maki RA: The macrophage and B cell-specific transcription factcor PU.1 is related to the ets oncogene. Cell 1990;61:113–124.
51 Goebl MG: The PU.1 transcription factor is the product of the putative oncogene *Spi-1*. Cell 1990;61:1165–1166.
52 Moreau-Gachelin F, Tavitian A, Tambourin P: *Spi-1* is a putative oncogene in virally induced murine erythroleukemias. Nature 1988;331:277–280.
53 Schuetze S, Stenberg PE, Kabat D: The Ets-related transcription factor PU-1 immortalizes erythroblasts. Mol Cell Biol 1993;13:5670–5678.
54 Galson DL, Hensold JO, Bishop TR, Schalling M, D'Andrea AD, Jones C, Auron PE, Housman DE: Mouse b-globin DNA-binding protein B1 is identical to a proto-oncogene, the transcription factor Spi-1/PU.1, and is restricted in expression to hematopoietic cells and the testis. Mol Cell Biol 1993;13:2929–2941.

55 Maxleod K, Leprince D, Stehelin D: The *ets* gene family. Trends Biochem Sci 1992;17:251–256.
56 Hromas R, Orazi A, Neiman RS, Maki R, Van Beveran C, Moore J, Klemsz M: Hematopoietic lineage- and stage-restricted expression of the ETS oncogene family member *PU.1*. Blood 1993;82:2998–3004.
57 Nelsen B, Tian G, Erman B, Gregoire J, Maki R, Graves B, Sen R: Regulation of lymphoid-specific immunoglobulin μ heavy chain gene enhancer by ETS domain proteins. Science 1993;261:82–86.
58 Pongubala JM, Nagulapalli S, Klemsz MJ, McKercher SR, Maki RA, Atchison ML: PU.1 recruits a second nuclear factor to a site important for immunoglobulin k 3′ enhancer activity. Mol Cell Biol 1992; 12:368–378.
59 Eisenbeis CF, Singh H, Storb U: PU.1 is a component of a multiprotein complex which binds an essential site in the murine immunoglobulin λ2–4 enhancer. Mol Cell Biol 1993;13:6452–6461.
60 Hagman J, Grosschedl R: An inhibitory carboxyl-terminal domain in Ets-1 and Ets-2 mediates differential binding of ETS family factors to promoter sequences of the *mb-1* gene. Proc Natl Acad Sci USA 1992; 89:8889–8893.
61 Feldhaus A, Mbangkollo D, Arvin KL, Klug CA, Singh H: BLyF, a novel cell-type- and stage-specific regulator of the B-lymphocyte gene. Mol Cell Biol 1992;12:1126–1133.
62 Shin MK, Koshland ME: Ets-related protein PU.1 regulates expression of the immunoglobulin J-chain gene through a novel Ets-binding element. Genes Dev 1993;7:20006–22015.
63 Zhang DE, Hetherington CJ, Chen JM, Tenen DG: The macrophage transcription factor PU.1 directs tissue-specific expression of the macrophage colony-stimulating factor receptor. Mol Cell Biol 1994;14: 373–381.
64 Pahl HL, scheibe RJ, Zhang DE, Chen HM, Galson DL, Maki RA, Tenen DG: The proto-oncogene PU.1 regulates expression of the myeloid-specific CD11b promoter. J Biol Chem 1993;268:5014–5020.
65 Mombaerts P, Iacomini J, Johnson RS, Herrup K, Tonegawa S, Papaioannou VE: RAG-1-deficient mice have no mature B and T lymphocytes. Cell 1992;68:869–877.
66 Shinaki Y, Rathbun G, Lam K, Oltz EM, Stewart V, Mendelson M, Charron J, Datta M, Young F, Stall AM, Alt FW: RAG-2-deficient mice lack mature lymphocytes owing to inability to initiate V(D)J rearrangement. Cell 1992;68:855–867.
67 Mucenski ML, Mclain K, Kier AB, Swerdlow SH, Schreiner CM, Miller TA, Pietryga DW, Scott WJ, Potter SS: A functional *c-myb* gene is required for normal murine fetal hepatic hematopoiesis. Cell 1991;65: 677–689.
68 Palacios R, Imhof BA: At day 8–8.5 of mouse development the yolk sac, not the embryo proper, has lymphoid precursor potential in vivo and in vitro. Proc Natl Acad Sci USA 1993;90:6581–6585.
69 Cumano A, Paige CJ, Iscove NN, Brady G: Bipotential precursors of B cells and macrophages in murine fetal liver. Nature 1992;356:612–615.
70 Hirayama F, Shih JP, Awgulewitsch A, Warr GW, Clark SC, Ogawa M: Clonal proliferation of murine lymphohemopoietic progenators in culture. Proc Natl Acad Sci USA 1992; 89:5907–5911.

Immunol Res 1994;13:291–298

Ursula Storb
Patricia Roth
Bradley Kurtz

Department of Molecular Genetics and Cell Biology, University of Chicago, Ill., USA

γ2b Transgenic Mice as a Model for the Role of Immunoglobulins in B Cell Development

Key Words
B lymphocyte development
Immunoglobulin gene expression
Feedback inhibition of heavy chain gene rearrangement
Transgenic mice with μ and/or γ2b transgenes
PreB cell receptor

Abstract
The development of B lymphocytes is tightly linked to the expression of immunoglobulins (Igs). Pro/preB cells which do not correctly rearrange heavy/light chain genes are aborted. Correctly rearranged Ig transgenes are apparently recognized by the developing B cells and can prevent the rearrangement of endogenous Ig genes. Both μ and γ2b heavy chain genes cause this feedback inhibition of heavy chain gene rearrangement. μ transgenes can in addition replace endogenous μ in its preB cell survival/maturation function. However, several different transgenic lines have shown that γ2b transgenes do not provide the nurturing functions of μ, except for one unique γ2b transgenic line, the C line. In this line mature B cells express γ2b only. Presumably, an unknown gene has been activated at the transgene integration site whose product overcomes the need for μ. The function of this gene depends of the presence of the surrogate light chain (sL), and thus must operate in combination with the preB cell receptor or in a downstream signaling/antiapoptosis event requiring the γ2b/sL receptor. The analysis of the two types of γ2b transgenic mice shows that the signals for preB cell development are highly complex and promises to reveal new insights into the molecular and cellular mechanisms of B cell maturation.

Introduction

The normal development of B lymphocytes has been shown to require the production of immunoglobulins (Igs) [1]. As preB cells mature, they first rearrange heavy chain D and J genes, later a V_H gene is joined to produce a VDJ_H combination. If this VDJ complex is in a correct translational reading frame to encode a functional heavy chain protein,

Ursula Storb
University of Chicago
Department of Molecular and Cell Biology
920 East 58th Street
Chicago, IL 60637 (USA)

the cell will proceed to the next stage, that of light chain gene rearrangement. If neither heavy chain gene allele is productively rearranged, the preB cell will die, apparently by apoptosis. The requirement for a productive heavy chain has been confirmed with the production of knockout mice lacking the exons encoding the transmembrane region of the μ gene (μmem) [2]. These mice completely lack B cells.

Mice with a deletion of μmem can be made to produce near normal B cell numbers, if a full μ transgene has been introduced by crossing with μ transgenic mice [3]. Futhermore, wild-type mice carrying a μ transgene have many transgenic μ^+ B cells in which the endogenous heavy chain genes are unrearranged or nonproductively rearranged [4–6]. Thus, the production of μ is required and sufficient for B cell development, and a single VDJ combination as provided by a μ transgene can replace the varied VDJ repertoire normally present during B cell development. Transgenic mice with a δ heavy chain transgene also show normal B cell development [7]. Many B cells in these mice produce only δ and show nonproductive endogenous μ genes. Thus δ can apparently substitute for μ.

It is not clear why in the μmem knockout mice a δ protein is not produced via RNA splicing of a VDJ to Cδ to make up for the lack of μ. Perhaps in early preB cells, when the production of μ first becomes critical, the δ locus is not in open chromatin (despite the accessibility of a transgenic δ locus) and transcription does not proceed that far 3′ from the μ locus. Alternatively, perhaps the manipulations of the membrane exons of μ have inactivated a control region necessary for transcription or mRNA splicing of VDJ to Cδ.

The μ and δ proteins are similar, but not identical in structure. Since both can permit early B cell development it was important to determine the structural components required for the nurturing function. It appears now that μ, and presumably δ, has several functions along the path of preB cell development. Besides being required for maturation of preB cells, μ synthesis in preB cells results in feedback inhibition of heavy chain gene rearrangement, as indicated above [4–6]. Furthermore, it seems to cause upregulation of light chain gene rearrangement [8]. Finally, μ, in combination with a functional light chain, is involved in the feedback inhibition of the V(D)J recombinase [reviewed in ref. 1]. It is not clear which structural portions of μ are responsible for these four different functions. Therefore it was of interest to produce transgenic mice with a γ2b transgene in order to determine if this related constant (C) region structure could substitute for the C region of μ.

γ2b Transgenes Prevent Normal B Cell Development

The γ2b transgene used in these experiments is shown in figure 1B. It represents a functional γ2b gene whose VH sequence is a member of the J558 family [9]. A portion of the JC intron is deleted, including most of the switch region. This deletion does not result in any detectable defect in transgene expression; although one would expect these γ2b transgenes not to switch their C region into an endogenous C region, because of the interference with the switch region, μ transgenes containing the complete JC intron have been shown to switch [10].

Several independent lines of γ2b transgenic mice were produced. All express high levels of γ2b mRNA in all lymphoid organs, including the thymus, but not in other tissues [9]. Expression in T cells is probably due to the presence of the heavy chain intron enhancer. Likewise, μ transgenes are expressed in T cells [6, 11]. In the transgenic T cells the respective heavy chains (γ2b or μ) are present in the cytoplasm, but they do not appear on the cell surface. For γ2b, as previously shown for μ [6, 11], both the membrane and secreted forms of γ2b mRNA are produced with a preponderance of the secreted form. The γ2b heavy chain is abundant among serum proteins of the transgenic mice.

Four of the five independent γ2b transgenic lines produced in our laboratory have a severe B cell defect [3, 12]. (The fifth line will be discussed in the next section.) A similar B cell defect seems also to exist in γ2b and γ1 transgenic mice produced in other laboratories [13–15]. At a young age, the γ2b mice have severely reduced numbers of B cells; only after about 6 months are nearly normal numbers of B cells present in the lymphoid organs. However, every B cell in these γ2b mice coexpresses μ derived from endogenous heavy chain gene rearrangement [3]. This is quite in contrast to μ transgenic mice which have generally only a mild B cell defect and many of whose B cells express the transgenic μ only [4–6, 16].

The B cell defect is unlikely due to toxicity of the γ2b since crossing the γ2b mice with μ transgenic mice alleviates the B cell defect [3]. Thus, the B cell defect in γ2b transgenic mice appears to be due to the lack of a function which μ possesses, but which γ2b lacks. In γ2b/μ double transgenic mice the B cells express both heavy chains. Interestingly, compared with the mice carrying only the μ transgene, the γ2b/μ double transgenic mice have a very strong suppression of endogenous μ production. This suggests that the γ2b transgene prevents endogenous heavy chain gene rearrangement, thus interfering with the production of a functional μ chain, but that it is unable to provide the maturation function.

The lack of a maturation function was shown directly by crossing the γ2b transgenic mice with μmem knockout mice [3]. The homozygous knockout mice carrying a γ2b transgene are completely B cell-deficient, similar to the knockout mice lacking a heavy chain transgene. Thus, γ2b cannot replace μ, whereas a μ transgene crossed into the μmem knockout mice completely restores B cell maturation [3].

The maturation defect with γ2b transgenes appears to block the transition of large, Ig⁻,

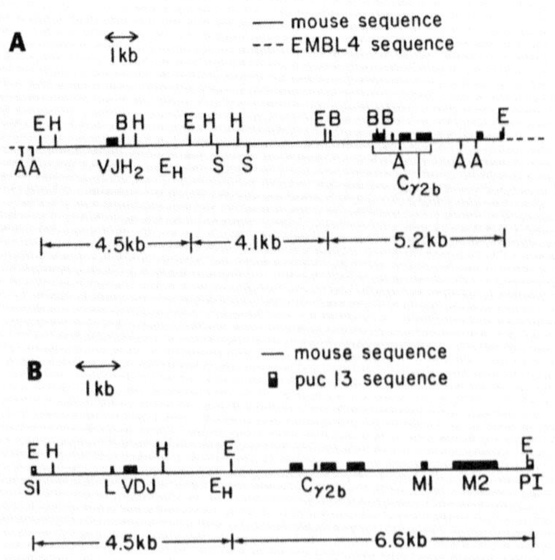

Fig. 1. Maps of γ2b gene [9]. **A** Original γ2b clone. The most 5′ and 3′ *Eco*RI sites are located in the EMBL4 vector DNA just next to the *Bam*HI cloning site. The black boxes indicate (from left to right): VDJ_H; Cγ2b, including C_{H1}, hinge, C_{H2}, C_{H3}, and introns; M1, the first transmembrane exon, just 3′ of the most 3′ *Ava*I site. **B** The γ2b transgene: this is derived from the DNA clone shown in **A** and a DNA clone containing the γ2b membrane exons. The *Sal*I and *Eco*RI sites at the left and *Eco*RI and *PVU*I sites at the right end, respectively are located within the pUC 13 vector. For production of the transgenic mice, the DNA was cut with *Sal*I and *Pvu*I and the resulting 11.2-kb fragment was isolated. L = Leader; VDJ = V, D, J region; EH = heavy chain enhancer; Cγ2b = C region of γ2b; M1 and M2 = membrane exons. A = *Ava*I, B = *Bam*HI, E = *Eco*RI, H = *Hin*dIII, PI = *Pvu*I, S = *SAc*I, S1 = *Sal*I.

	ProB	Early preB	Late preB	B
Normal	DJ ⟶ VDJCμ			
		μ ⟶ inhibits further VD rearrangement		
		μ ⟶ promotes B cell development ⟶		μ + B cells
γ2b Transgenic				
Most preB	γ2b ON early: γ2b does not foster B cell development ⟶ no			γ-only B cells
	γ2b prevents VD rearrangement: no μ production			
Rare preB	DJ ⟶ VDJCμ			
		μ ⟶ promotes B cell development ⟶		μ + γ + B cells
		γ2b ON late: γ does not inhibit B cell development		
'C' γ2b Transgenic				
	γ2b ON: active integration site promotes B cell development ⟶			γ-only B cells
	γ2b prevents VD rearrangement: no μ			

Fig. 2. Model of B cell development in two types of γ2b transgenic mice [21]. See text.

B220$^+$ preB cells into small Ig$^-$, B220$^+$ preB cells [3, 17]. This is the same stage at which mice with a knockout of the λ5 of the μmem gene show a developmental defect [18].

The results suggest a model of B cell development in γ2b transgenic mice shown in figure 2 (central panel). Early preB cells produce sufficient quantities of transgenic γ2b to block the rearrangement of endogenous heavy chain genes. However, γ2b cannot provide a survival or maturation function which is required for further B cell development. Perhaps, the arrested large preB cells die by apoptosis in analogy to findings with mice which carry the *scid* defect [19]. Thus, the production of B cells is essentially blocked. In very rare preB cells, the synthesis of transgenic γ2b may be delayed, or the rearrangement of endogenous heavy chain genes hastened, so that μ precedes the γ2b production. If the rearrangement event is productive, such preB cells will produce μ and thus be provided the nurturing stimulus. Mature B cells arising from such an event coproduce γ2b with endogenous μ. If these rarely arising B cells are long-lived, they will accumulate in the periphery. Thus, old γ2b transgenic mice have almost normal numbers of B cells, all of which produce endogenous μ.

It is not clear at this time, how μ can provide a nurturing stimulus which γ2b cannot provide, and which part of the μ molecule is responsible for the effect. Transgenic mice with a γ2b/μ hybrid transgene in which the transmembrane and intracytoplasmic portions of γ2b have been replaced by those of μ are as B cell-deficient as the conventional γ2b transgenic mice and show feedback inhibition of heavy chain gene rearrangement [3]. The transmembrane portion of μ is similar to that of γ2b (and δ), but the intracytoplasmic portion is very different. Thus, the transmembrane portion of γ2b may have the feedback effect, but that portion of μ combined with the short μ intracytoplasmic tail is not sufficient to provide the maturation function. As summarized above, μ seems to have four functions during early B cell development. Clearly, the feedback inhibition of heavy chain gene rearrangement is shared by γ2b. It also appears likely that γ2b has the light chain rearrangements upregulating effect: The quantities of κ mRNA are the same in fetal liver and bone marrow of γ2b transgenic and normal mice [17]. Futhermore, when γ2b transgenic mice are crossed with λ light chain transgenic mice, the defect is not alleviated [3, 17]. This particular λ transgene is under the control of the heavy chain enhancer [20] and is therefore expressed early in B cell development. Thus light chains should be present at the normal time, if not earlier, and if the nurturing μ effect were due to light chain upregulation only, the λ transgene should have overcome the need for this function of μ.

It is not clear if γ2b, in combination with light chains, can exert the recombinase feedback effect, however, data described in the next section suggest that this may be so. Besides, lack of this late function of μ cannot be responsible for the block at the early preB cell stage seen in γ2b transgenic mice. Thus the only remaining function of μ which is deficient when γ2b replaces μ seems to be the ill-defined survival/maturation effect. This possibility will be discussed in the next section.

A Unique γ2b Transgenic Mouse Is Capable of Normal B Cell Development

All γ2b or γ1 transgenic mice produced in our or other laboratories show the described phenotype, except one unique mouse line, named the C line [21]. This mouse expresses the γ2b transgene in B and T cells at similar levels as the other γ2b transgenic mice. However, most of the B cells produce γ2b only and there does not seem to be a need for μ in this line. This unusual mouse line is of interest, because, as discussed below, it will serve as a model for the further unraveling of the events which regulate early B cell development.

In contrast to the other γ2b transgenic lines, in the C line the distribution of B cell precursors in the bone marrow and periphery is essentially normal. There is a slight decrease in total B cells similar to that seen with μ transgenes and perhaps related to the fact that most or all of the cells express a single heavy chain, thus greatly reducing the B cell receptor repertoire.

When the C line is crossed with a conventional γ2b transgenic line, the C phenotype is dominant: the crosses have normal B cell numbers and the B cells express only γ2b [21].

The unique phenotype of the C line is not due to an inability to express μ. When the C mice are crossed with a μ transgenic line, the B cells coexpress the transgenic μ with γ2b [21]. These crosses with μ transgenics also show an almost complete suppression of endogenous μ gene expression, just like the crosses of the conventional γ2b-transgenic mice with μ transgenics described above.

The data suggested that feedback inhibition of endogenous heavy chain gene rearrangement is the same in the C line and other γ2b transgenics, but that in the C line μ is not required for B cell development (fig. 2). This was shown directly by crossing mice of the C line with a μmem knockout line: such crosses have essentially normal numbers of B cells which of course express only γ2b [21]. In contrast, as described above, such crosses with conventional γ2b transgenic mice are completely B cell-deficient.

What is the molecular basis for the unique behavior of the C line? The effect cannot be due to inactivation of an endogenous locus, since the C phenotype is dominant and the transgene has always been transmitted only in hemizygous form. Inactivation of the wild-type allele by parental imprinting is ruled out, because the transgene has been transmitted through males and females with equal effect.

Superactivation of transgene expression by a position effect is also unlikely, because the levels of the membrane and secreted forms of

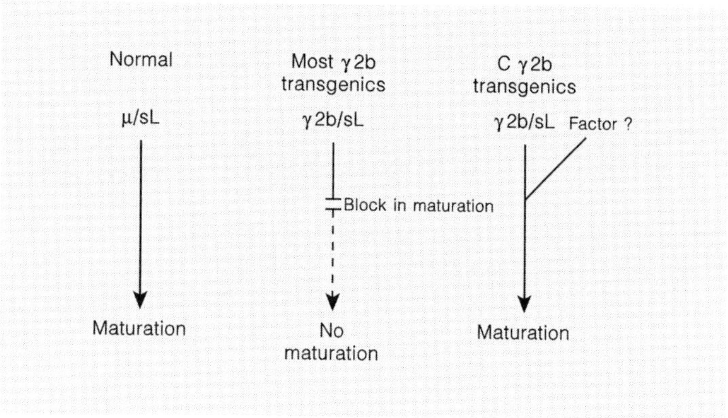

Fig. 3. Scheme of signaling via different preB cell receptors [21]. An unknown factor makes a γ2b/sL receptor signal functionally equivalent to a μ/sL receptor signal.

the γ2b mRNA are similar in all γ2b transgenic mice. Furthermore, the C transgene is expressed slightly later in B cell development than the transgenic mRNA in other γ2b transgenic mice. However, in crosses between C and other γ2b transgenics the γ2b expression is earlier, as in the other γ2b transgenics, but the phenotype is that of the C mouse. Thus, the small differences in the timing of transcription cannot be responsible for the C line effect.

Since the C line can dispense with the need for μ, the possibility existed that one of the three transgene copies in this line may have been mutated during the transgene integration to resemble the μ structure and therefore replace the μ function. However, cloning and sequencing of all three transgene copies revealed no difference from the transgenes of the other γ2b transgenic mice studied in our laboratory.

Given these considerations, the most likely possibility is that in the C line the transgenes are integrated at a position where their presence in preB cells activates an endogenous gene whose expression obviates the need for μ. As discussed in the section concerning the conventional γ2b transgenic mice, the role of μ which the conventional γ2b transgenes cannot replace is most likely the survival/maturation function. Results with the C mouse further support the idea that γ2b is not deficient in the upregulation of light chain gene expression. The gene postulated to be activated at the integration site cannot be a light chain gene since the light chains reside on chromosomes 6 and 16, but the transgene integration in the C mouse is in chromosome 13 [B. Kurtz, R. Espinoza, M. LeBeau and U. Storb, unpublished].

The replacement of the survival/maturation function of μ presumably indicates that a cellular signal is created that circumvents the need for μ (fig. 3). The signals generated by μ apparently operate through the preB cell receptor which is composed of μ and the surrogate light chains (sL), λ5 and VpreB [18]. The possibility existed that in the C mouse a gene was activated in preB cells which circumvents μ by interphasing with a point in a signal cascade downstream of the preB cell receptor. In that case, sL would not be required. Since it was not known if γ2b can associate with sL, γ2b was transfected into an sL^+ preB cell line. Transfectants showed association of γ2b with sL. Mice of the C line (and as control a conventional γ2b line) were then crossed with a mouse with a deletion of the λ5 genes [18].

Such crosses were completely B cell-deficient, indicating that in the C mouse a signal involving the sL is required for B cell maturation [21]. This also indirectly confirms that γ2b can can associate with sL.

Thus, the function of the postulated gene activated at the transgene integration site in the C line is either coincident with the γ2b/sL signal or is a downstream event within the preB cell. The signals via the preB cell receptor are not well understood, but can be assumed to be complex, given the multiple signals created by activation of the mature B cell receptor [22]. In the C line a protein may be recruited which, in combination with γ2b/sL, results in a μ-like, complex receptor signal. Alternatively, a ligand for the preB cell receptor that cannot be recognized by the γ2b/sL may be constitutively firing in the C line. Other cells interacting with preB cells are stromal cells and perhaps T cells. The unknown gene is unlikely to be a stromal cell gene, because transfer of C bone marrow to irradiated normal mice results in a C phenotype [P.R. and U.S., unpublished]. Stromal cells are not expected to survive the transfer protocol.

T cells are not normally required for early B cell development. When mice with deletions of the RAG-1 or RAG-2 genes are provided with transgenic heavy and light chain genes, they develop apparently normal B cells [23, 24]. Such RAG knockout mice have no T cells. However, it is possible that in the C line a novel T cell-preB cell interaction is constitutively active and overcomes the need for μ. This is being experimentally tested.

Alternatively to acting coincident with the preB cell receptor, the postulated function in the C line may be a downstream function normally induced by μ/sL, but not inducible by γ2b/sL. It seems likely, as discussed earlier, that the γ2b/sL receptor only provides a subset of signals generated by μ/sL (fig. 3). It is also possible that the activated gene does not normally function during B cell development, but can substitute for the missing signal when ectopically expressed. In any case, further investigation of the C mice may reveal stages and molecular interactions in the receptor signaling cascade in preB cells.

Finally, apoptosis has to be considered. Presumably, preB cells which do not receive a proper survival/maturation signal die by apoptosis [25, 26]. It is possible that in the C mouse, an antiapoptosis gene has been induced constitutively, or is induced after triggering via the γ2b/sL recpetor.

Whatever the alteration in the C line, it cannot operate independently of the γ2b/sL receptor (fig. 3). These data clearly show that the mechanisms of preB cell development are very complex, requiring a functional heavy chain together with other signals. The two types of γ2b transgenic mice will be useful in the dissection of the molecular and cellular components of early preB cell development.

Acknowledgements

Work from our laboratory cited herein was supported by NIH grants HD23089 and AI24780. We are grateful to Emily Klotz for critical reading of this paper.

References

1 Storb U: Ig gene expression and regulation in Ig transgenic mice; in Alt F, Honjo T (eds): Immunoglobulin Genes. 1994.

2 Kitamura D, Roes J, Kuhn R, Rajewsky K: A B cell-deficient mouse by targeted disruption of the membrane exon of the immunoglobulin μ chain gene. Nature 1991;350:423–426.

3 Roth P, Doglio L, Manz J, Kim JY, Lo D, Storb U: Immunoglobulin γ2b transgenes inhibit heavy chain gene rearrangement, but cannot promote B cell development. J Exp Med 1993;178:2007–2021.

4 Manz J, Denis K, Witte O, Brinster R, Storb U: Feedback inhibition of immunoglobulin gene rearrangement by membrane μ but not by secreted μ heavy chains. J Exp Med 1988;168:1363–1381.

5 Rusconi S, Kohler G: Transmission and expression of a specific pair of rearranged immunoglobulin μ and κ genes in a transgenic mouse line. Nature 1985;314:330–334.

6 Weaver D, Costantini F, Imanishi-Kari T, Baltimore D: A transgenic immunoglobulin mu gene prevents rearrangements of endogenous genes. Cell 1985;42:117–127.

7 Iglesias A, Lamers M, Kohler G: Expression of immunoglobulin delta chain causes allelic exclusion in transgenic mice. Nature 1987;330:482–484.

8 Ehlich A, Schaal S, Gu H, Kitamura D, Muller W, Rajewsky K: Immunoglobulin heavy and light chain genes rearrange independently at early stages of B cell development. Cell 1993;72:695.

9 Tsang H, Pinkert C, Hagman J, Lostrum M, Brinster RL, Storb U: Cloning of a γ2b gene encoding anti-*Pseudomonas aeruginosa* H chains and its introduction into the germ line of mice. J Immunol 1988;141:308–314.

10 Gerstein R, Frankel W, Hsieh CL, Durdik JM, Rath S, Coffin JM, Nisonoff A, Selsing E: Isotype switching of an immunoglobulin heavy chain transgene occurs by DNA recombination between different chromosomes. Cell 1990;63:537–548.

11 Storb U, Pinkert C, Arp P, Engler P, Gollahon K, Manz J, Brady W, Brinster RL: Transgenic mice with μ and κ genes encoding antiphosphocholine antibodies. J Exp Med 1986;164:627–641.

12 Denis K, Provost S, Witte O, Brinster R, Storb U: Delay of early B-lymphocyte development by gamma 2b immunoglobulin transgene: Effect on differentiation-specific molecules. Dev Immunol 1990;1:105–112.

13 Offen D, Spatz L, Escowitz H, Factor S, Diamond B: Induction of tolerance to an IgG autoantibody. Proc Natl Acad Sci USA 1992;89:8332–8336.

14 Tsao BP, Ohnishi K, Cheroutre H, Mitchell B, Teitell M, Mixter P, Kronenberg M, Hahn BH: Failed self-tolerance and autoimmunity in IgG anti-DNA transgenic mice. J Immunol 1992;149:350–358.

15 Yamamura K-I, Kudo A, Ebihara T, Kamino K, Araki K, Kumahara Y, Watanabe T: Cell-type-specific and regulated expression of a human γ1 heavy-chain immunoglobulin gene in transgenic mice. Proc Natl Acad Sci USA 1986;83:2152–2156.

16 Kenny JJ, Finkelman F, Macchiarini F, Kopp WC, Storb U, Longo DL: Alteration of the B cell surface phenotype immune response to phosphocholine and the B cell repertoire in M167 μ plus κ transgenic mice. J Exp Med 1989;142:4466–4474.

17 Roth P: Characterization of γ2b Transgenic Mice: Effects on B Cell Development and Allelic Exclusion; doctoral thesis University of Chicago, 1993.

18 Kitamura D, Kudo A, Schaal S, Mueller W, Melchers F, Rajewsky K: A critical role of λ5 protein in B cell development. Cell 1992;69:823–831.

19 Strasser A, Harris A, Corcoran L, Cory S: Bcl-2 expression promotes B- but not T-lymphoid development in *scid* mice. Nature 1994;368:457–460.

20 Hagman J, Lo D, Doglio LT, Hackett J Jr, Rudin CM, Haasch D, Brinster R, Storb U: Inhibition of immunoglobulin gene rearrangement by the expression of a λ2 transgene. J Exp Med 1989;169:1911–1929.

21 Roth P, Kurtz B, Lo D, Storb U: λ5, but not μ is required for B cell maturation in a unique γ2b transgenic mouse line. J Exp Med 1995, in press.

22 Gold M, DeFranco A: Biochemistry of B lymphocyte activation. Adv Immunol 1994;55:221–295.

23 Spanopoulou E, Roman C, Corcoran L, Schlissel MS, Silver DP, Nemazee D, Nussenzweig MC, Shinton SA, Hardy RR, Baltimore D: Functional immunoglobulin transgenes guide ordered B-cell differentiation in Rag-1-deficient mice. Genes Developm 1994;8:1030–1042.

24 Young F, Ardman B, Shinkai Y, Lansford R, Blackwell TK, Mendelsohn M, Rolink A, Melchers F, Alt FW: Influence of immunoglobulin heavy- and light-chain expression on B cell-differentiation. Genes Developm 1994;8:1043–1057.

25 Osmond DG: Proliferation kinetics and the lifespan of B cells in central and peripheral lymphoid organs. Curr Opin Immunol 1991;3:179–185.

26 Rolink A, Karasuyama H, Haasner D, Grawunder U, Martensson I, Kudo A, Melchers F: Two pathways of lymphocyte development in mouse bone marrow and the roles of surrogate L chain in this development. Imm Rev 1994;137:185–201.

Immunol Res 1994;13:299–310

Jeannie Tseng
Young Jae Lee
Bartholomew J. Eisfelder
Marcus R. Clark

Departments of Medicine and Pathology and Committee on Immunology, University of Chicago, Chicago, Ill., USA

The B Cell Antigen Receptor Complex: Mechanisms and Implications of Tyrosine Kinase Activation

Key Words
B cell antigen receptor
Tyrosine kinases
Immunoglobulin-α
Immunoglobulin-β
CD45

Abstract
The B cell receptor is a multimeric receptor complex whose constituent chains appear to mediate distinct and possibly interrelated functions. In this review we have focused on how one chain, immunoglobulin (Ig)-α, may function to activate tyrosine kinases and the consequences of that activation. The cytoplasmic domain of Ig-α contains a consensus sequence, the antigen recognition homology 1 (ARH 1) motif, which is found in Ig-β and other antigen recognition receptor associated chains. We argue that this conserved structure reflects an underlying conserved mechanism of secondary effector activation. Our data also indicates that the specificity of each motif (i.e., the elements which restrict secondary effector binding to particular motifs) is encoded within divergent sequences found in each ARH 1 motif. In the particular case of kinase activation by Ig-α, the subsequent phosphorylation of multiple tyrosines on Ig-α, Ig-β, CD19, CD22 and possibly other functionally related chains form recruitment sites for a myriad of secondary signal transducers. In this model, proximal tyrosine kinases and phosphatases do not function so much to mediate the linear transfer of information as to establish and modulate an interrelated network of signal transducers capable of driving complicated cellular responses.

Introduction

A B cell's recognition of, and response to, native antigen is dependent upon the surface expression of a clonotypic, antigen-specific, B cell receptor (BCR). Depending on the developmental stage of the cell and the context in which stimulation occurs, receptor engagement can elicit a myriad of phenotypic changes including cell activation, differentiation, anergy, or cell death [1]. These divergent biological responses are triggered by a BCR-initiated cascade of intracellular events involving, most proximally, the activation and/or recruitment of tyrosine kinase(s) [2–4]. These kinases, in turn, stimulate at least three interrelated pathways of signal transduction. Within 1 min of stimulation, both phospholi-

Supported by grants from the NIH (AR01864) and Arthritis Foundation.

Dr. Marcus R. Clark
Departments of Medicine and Pathology
Committee on Immunology, University of Chicago
5841 South Maryland Avenue, MC0930
Chicago, IL 60637 (USA)

pase Cγ1 and 2 (PLCγ) [5, 6] are phosphorylated and activated, thereby initiating calcium mobilization and protein kinase C activation. Upon antigen receptor stimulation, ras forms an activation complex with Shc, Grb2 and mSOS which is competent to recruit Raf to the plasma membrane and initiate a cascade of serine/threonine kinase activation necessary for proliferation [7–12]. Alternative mechanisms of ras activation involving the augmentation of the nucleotide exchange activity of VAV [13, 14] or attenuation of the activity of ras GTPase activating protein may also occur [15]. The relative importance of these latter mechanisms of B cell antigen receptor-mediated ras activation is unknown. Finally, phosphatidylinositol 3 kinase, which may be activated by direct interaction with Src-family kinases [16] or with ras [17], appears to initiate another cascade of activation which has been shown in other systems to be necessary for proliferation and differentiation [18, 19]. Other mediators of signal transduction have been found in proximity to the BCR including a pertussis toxin-sensitive GTP-binding protein [20] and other unidentified serine/threonine kinases [21; Clark et al., unpublished observation].

Thus, the BCR is a biologically complex, plastic and contextually dependent structure which utilizes a variety of signal transduction pathways to drive divergent cellular processes. This functional complexity is in contrast to the original description of the resting B cell antigen receptor as having two cytoplasmic tails of three amino acids each [22, 23]. Such a small cytoplasmic domain seemed insufficient to couple the receptor to intracellular signaling pathways. However, the subsequent biochemical elucidation of a multimeric B cell antigen receptor complex revealed a structure capable of initiating antigen-driven B cell activation.

Structure of the Receptor Complex

When the antigen receptor was precipitated from digitonin lysates of B lymphocytes it became clear that membrane immunoglobulin (Ig) did not exist in isolation, but rather as part of a complex containing disulfide-linked heterodimeric glycoproteins which were inducibly tyrosine phosphorylated upon membrane Ig cross-linking [24–27]. These molecules, Ig-α and Ig-β, were subsequently shown to be the respective products of the mb-1 and B29 genes [28–31]. These polypeptides contain an Ig-like extracellular domain, a transmembrane spanning region and significantly, extensive cytoplasmic domains of 48 amino acids (Ig-β) and 61 amino acids (Ig-α) [32, 33]. The heterodimer which is necessary for surface expression of membrane Ig [34] is thought to interact with a polar patch in the transmembrane spanning region of μ heavy chain (TAST). In the absence of Ig-α/β, this patch signals retention of the heavy chain within the endoplasmic reticulum [1, 35, 36]. Other sites of interaction between Ig-α/β and heavy chain have been proposed, although the overall structure and stoichiometry of the complex are unknown.

Like the BCR, the T cell antigen receptor is assembled and expressed as a multimeric complex containing invariant chains [37]. The antigen binding substructure is coexpressed with CD3, a complex of three homologous heterodimeric transmembrane proteins – the ε, γ and δ chains and disulfide-linked homo- or heterodimers of ζ and η chains. The inclusion of these invariant dimeric complexes in receptors capable of cellular activation suggested that they were vested with signaling abilities. This possibility was first tested by Irving and Weiss [38], who constructed chimeric proteins linking the cytoplasmic domains of the T cell receptor (TCR) ζ chain with the extracellular and transmembrane domains of CD8. Cross-linking the CD8/ζ chimeras activated tyrosine kinases and stimulated interleukin-2 production to levels comparable to those induced via the endogenous TCR. Similar studies were done with Ig-α/β. Chimeric cell surface receptors were created by linking the intracytoplasmic domains of Ig-α and Ig-β with the extracellular and transmembrane domains of CD8, IgM or FcγRII [39–43]. These chimeric proteins were expressed independently of the endogenous BCR, cross-linked and shown to be competent to mediate activation. These and other studies established that the invariant chains of these receptor complexes couple antigen recognition to intracellular signaling pathways.

In addition to the antigen receptors, other complex immune recognition receptors, such as hFcγRIII and FcεRI, have chains functionally and structurally homologous to Ig-α and Ig-β. When the primary amino acid structures of these chains are compared, a consensus sequence [$(D/E)X_7(D/E)XXYXXL(X)_{6-8}Y$ XXL/I] becomes apparent (fig. 1) [44–46]. This cytoplasmic sequence motif has since been termed the antigen recognition activation motif (ARAM) [37] or antigen recognition homology 1 (ARH 1) motif [47]. Extensive analysis of many of these motif-containing sequences has demonstrated that they encode for most, if not all, of the signal-transducing capacity of the parent chain in which they are embedded [48–51].

Ig-α and Ig-β Contribute Differential Functional Properties to the B Cell Antigen Receptor Complex

The appearance of this structural motif within the cytoplasmic domains of multiple chains found associated with the antigen receptors suggested that they may serve redundant functions within each complex and that

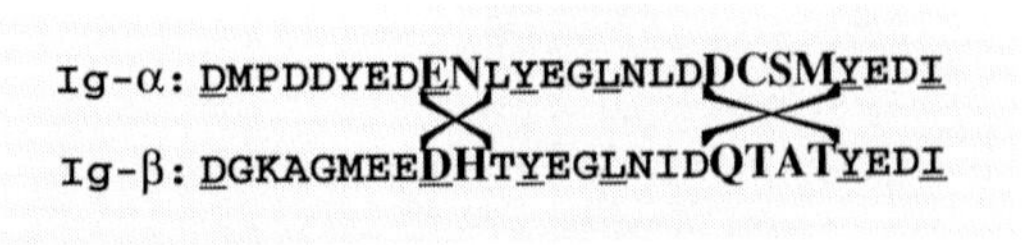

Fig. 1. Diagrammatic representation of the Ig-α and Ig-β ARH 1 motifs. Underlined are the conserved amino acid residues of the ARH 1 motif. Also indicated are the residues which were exchanged to determine that the binding specificity of Ig-α and Ig-β.

their multiplicity may serve to amplify weak signals. This view is supported by the observation that tandem repeats of the ARH 1 motif in the TCR ζ chain are more competent to transduce a signal than a single copy [52]. Furthermore, certain conserved residues, most importantly the two tyrosines of the ARH 1 motif, are necessary for signal transduction [48, 50, 53]. Again, this suggests that the motifs share a common functionality.

However, several experiments suggest that different ARH 1 motifs differ in their ability to activate specific receptor-associated functions. For example, an overexpressed TCRζ, but not TCRη, transgene is capable of causing premature termination of RAG-1 and RAG-2 expression [54] and subsequently, a block in thymocyte maturation prior to the appearence of $CD4^+/CD8^+$ cells. These transgenic studies suggest that the carboxy-terminal ARH 1 motif of TCRζ, which is not found in TCRη, possesses unique signal transducing functions which give rise to the disparate results observed. The invariant chains of the FcεRI receptor can also mediate different functions. When the β and γ cytoplasmic tails, each containing one ARH 1 motif, are expressed as chimeras, it is clear that the γ but not β can mediate serotonin release in basophils [55]. This functional difference may be due to the differential abilities of each of these motifs to interact with and activate the tyrosine kinases, Syk and Lyn.

In recent studies using chimeric receptors, evidence has been provided for the segregation of function within the Ig-α/Ig-β heterodimer. The cytoplasmic domain of Ig-α has been shown to fully restore intracellular Ca^{2+} mobilization, tyrosine phosphorylation and cytokine production [39–43] while Ig-β, in most instances, cannot. Activation by Ig-β, when it was observed, occurred much later than that induced by either the intact multimeric receptor complex or Ig-α alone and in most cases, the degree of activation was significantly less [39, 42]. These data suggest that within the Ig-α/Ig-β heterodimer, each subunit may contribute differential signaling properties to the antigen receptor complex despite sharing the same structural motif.

Therefore, there appears to be a certain duality of ARH 1 motif function. Structural comparison of the motifs as well as mutagenesis of conserved residues suggest that common mechanisms underlie secondary mediator binding/activation. However, divergent elements within each motif are likely to specify which effector molecules are able to interact with particular motifs.

Divergent Sequences within Ig-α and Ig-β Impart Functional Specificity

To begin to define the molecular basis for this duality of ARH 1 motif function, we undertook a structural analysis and comparison of Ig-α and Ig-β. From the studies described above, it was clear that truncations of single cytoplasmic domains containing one ARH 1 motif were capable of mediating many of the functions associated with the parent multimeric receptor complex. We postulated that this functionality required the ARH 1-containing chains to stably interact with secondary signal transduction mediators. Therefore, the cytoplasmic tails of Ig-α or Ig-β were expressed as fusions of glutathione S-transferase and were used to probe B cell lysates for binding proteins. These studies revealed that Ig-α and Ig-β ARH 1 containing cytoplasmic domains bind distinct sets of effector molecules. The Ig-α cytoplasmic tails bind the Src-family kinases Lyn, Fyn and Blk, and an unidentified protein of 38 kD. Consistent with the differing signaling properties of the above-described Ig-α and Ig-β chimeras, Ig-β cytoplasmic tails do not bind Src-family kinases efficiently. Instead Ig-β binds two unidentified proteins of 40 and 42 kD [47]. An extensive mutational analysis of the conserved tyrosines of each ARH 1 motif demonstrated that binding of the above proteins was independent of these tyrosines and thus could not depend on their phosphorylation [56].

The phosphotyrosine independence of these interactions is consistent with the experiments of Burkhardt et al. [3], in which multiple Src-family kinases were found in association with resting and presumably non-tyrosine-phosphorylated receptor complexes. These associations may determine the phenotype of each motif in that only those kinases and other effectors that are constitutively associated with the resting receptor can be immediately activated. In other words, the constituents of the resting receptor may determine which pathways can be utilized upon activation.

We reasoned that these important differences in binding activity of Ig-α and Ig-β may be determined by divergent sequences within each ARH 1 motif. Comparison of the sequence of the motifs revealed two regions of significant disparity flanking the conserved tyrosines (fig. 1), which could provide the basis for differential binding. A series of switch mutants between Ig-α and Ig-β determined that an amino acid sequence, DCSM, found uniquely in Ig-α determined specificity for p38 and Src-family kinase binding. Expression of this DCSM in the place of a QTAT sequence, which is found exclusively in Ig-β, was found to confer Ig-α-like properties to Ig-β, including Src-family kinase and p38 binding. Conversely, the Ig-β QTAT sequence determined p40 and p42 binding and upon generation of the switch mutants could endow this binding activity on Ig-α [56]. These data suggest that the binding specificity, but not necessarily the binding activity of each motif, is determined by divergent sequences located between the two conserved tyrosines. Recent evidence also predicts that these Src-family kinases constitutively associate with Ig-α via their N-terminal 10 amino acids [57].

A Common Mechanism of Activation

If the divergent sequences within each ARH 1 motif determine the specificity of each chain, what is the function of those residues which the chains have in common? One of the most immediate consequences of antigen receptor ligation is phosphorylation of the ARH 1 tyrosines of Ig-α and Ig-β [24]. Recent findings by Songyang et al. [58], which indicate that phosphorylated YEDI is a consensus binding site for the SH2 domains of Src-family kinases, raise the possibility that phosphorylation of Ig-α might change its binding to these molecules. To address this possibility, we constructed ARH 1 phosphopeptides and assessed their binding activity. Indeed, as defined by immunoblotting, Ig-α ARH 1, phosphorylated singly at YEGLN or YEDI, exhibited a 2- to 3-fold increase in Fyn binding activity while doubly phosphorylated Ig-α ARH 1 exhibited a ~20-fold increase in binding activity [56]. This synergy in the effects of tyrosine phosphorylation suggests cooperativity in effector binding. Furthermore, the binding of phosphorylated Ig-α to Fyn increased the activity of the enzyme 3-fold. These studies reveal a novel mechanism by which signals are propagated following the initial ligand-induced Ig-α tyrosine phosphorylation. Although the basis of this activation is unknown, data are most consistent with a derepression mechanism [59, 60], that is, upon binding of the Fyn SH2 domain to phosphorylated ARH 1, the stronger affinity of this interaction displaces the previously bound regulatory carboxy-terminal phosphotyrosine. This would then allow the kinase to unfold into an active conformation (fig. 2).

From our studies, it is unclear however why Fyn, with one SH2 domain, would require a doubly phosphorylated ARH 1 peptide for binding. Three possible explanations exist. One possibility is that an intermediate molecule or linker is necessary for Fyn to bind Ig-α. If such a molecule had two SH2 domains, then Fyn binding might require two phosphotyrosines. Binding would be dependent not only on tyrosine phosphorylation of Ig-α, but on the phosphorylation of this putative linker molecule. However, Ig-α can bind Fyn from the lysates of bacteria overexpressing Fyn, suggesting that Fyn can bind Ig-α directly without the need of an intermediate linker molecule (unpublished observation) [57]. Another possibility is that the SH2 domain of Fyn recognizes only one phosphotyrosine. In this case, phosphorylation at the other tyrosine would regulate the accessibility of the binding phosphotyrosine to Fyn. From an analysis of SH2 binding affinities, it is clear that Fyn should bind the YEDI motif with a higher affinity than the YEGL motif [58]. Therefore, the YEGL sequence may regulate the accessibility of the YEDI putative binding site. The hypothesis that the Ig-α ARH 1 motif consists of effector binding and regulatory subdomains is readily testable.

The crystalline structure of Lck suggests a third possibility [60]. It appears that Lck, and presumably other Src-family kinases, may ex-

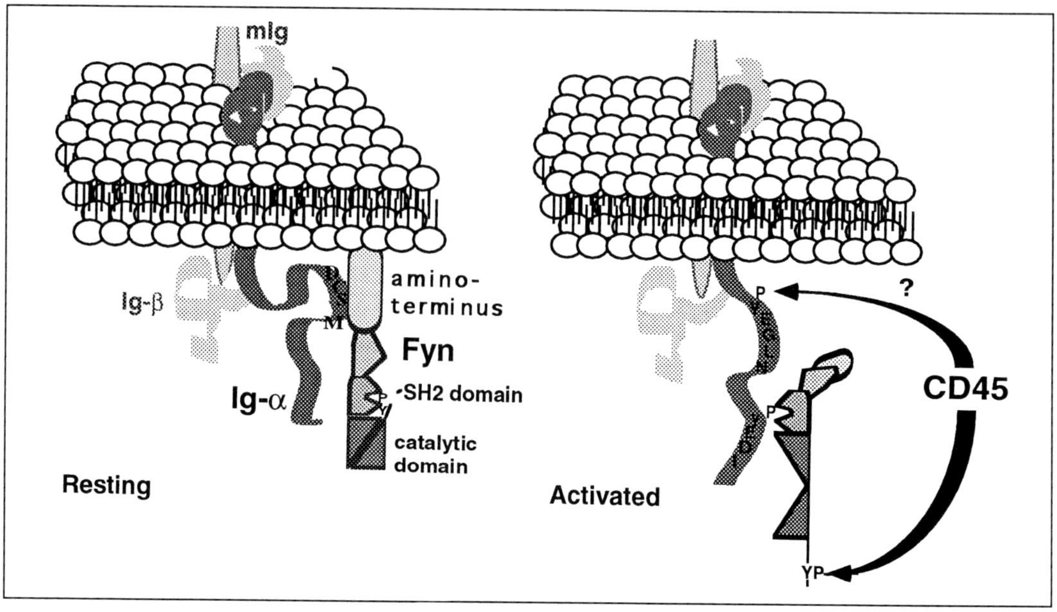

Fig. 2. Model of receptor mediated kinase activation. In the resting receptor complex, Fyn and other Src-family members are associated with Ig-α in an inactive state. Following receptor activation, phospho-Ig-α is postulated to act as a catalyst which facilitates the unfolding of Fyn, allowing the tyrosine phosphatase CD45 access to the negative regulatory phosphorylation site found on the kinase. CD45 may also dephosphorylate Ig-α which would allow the active kinase to dissociate from the receptor and interact with secondary effectors and/or substrates.

ist in the inactive state as homodimers formed by the SH3 domains of each kinase binding to carboxy-terminal proline-rich regions on the other. The carboxy-terminal phosphotyrosines of each kinase are bound to the SH2 domain of the same kinase. It is possible that the two phosphotyrosines of Ig-α must be present and available for binding both SH2 domains present in the dimer before the complex can be disrupted and efficient binding to Ig-α occurs. Interestingly this hypothesis predicts that mutations in the SH3 domain of Lck would enhance SH2 binding to singly phosphorylated ARH 1 motifs. To our knowledge, this has not been directly examined.

The Initiation of Signal Transduction

Our studies have focused on how a model Src-family kinase, Fyn, is constitutively associated with and activated by Ig-α. Implicit in these experiments is the assumption that members of the Src family of kinases are the primary kinases activated by the receptor complex. In support of this are the observations by Yamanashi et al. [61] and Burkhardt et al. [3] that multiple Src-family kinases are in close or direct association with the resting receptor complex and are activated within a few seconds of receptor pertubation. However another family of kinases (whose members include Syk and ZAP-70) has been implicated in antigen receptor-mediated signal transduction. In contrast to the Src-family members, these kinases contain two SH2 domains and no myristylation site (SH4 domain). The expression of ZAP-70 is restricted to T cells while Syk is found more widely in B cells, T cells and basophils. Initial studies which were focused on ZAP-70 suggested that this kinase functioned downstream of the Src-family kinases in TCR activation. In particular, chimeras of CD3ε or TCRζ can coimmunoprecipitate with ZAP-70 only after the receptor chains have been phosphorylated by Fyn or Lck [62, 63]. This recruitment appears to be mediated by the physical association of both SH2 domains of ZAP-70 with the tyrosine phosphorylated receptor chains [62, 64]. Although in this system, Lck clearly mediates the recruitment of ZAP-70 to the TCR, the kinases may also interact directly. In particular, Lck may phosphorylate and bind to ZAP-70 via its SH2 domain. This may enhance the specific activity of both kinases.

Recent observations suggest, however, that Syk may function differently. This kinase can be coprecipitated with the resting BCR and one group has recently provided evidence that Syk may be an immediate kinase effector, both upstream and independent of Src-family kinases. In an Lck–/– T cell line, JCaM1, they demonstrated that the association of Syk, but not ZAP-70, occurred independently of Lck [65]. Furthermore, the specific activity of Lck but not Syk was enhanced by coexpression of the two kinases [66]. However, it is unlikely that Syk activation is completely independent of Src-family kinases in that its phosphorylation following BCR stimulation is significantly decreased in a Lyn–/– B cell line as compared to that induced in Lyn+/+ cells. Thus, it is still unresolved whether Syk, the Src-family members or both mediate the first tyrosine phosphorylation events.

In any case, we propose that the resting BCR is constitutively associated with inactive tyrosine kinases. In the particular case of the Src-family kinases, we and others have shown that they associate, via their amino terminus, with the DCSM motif in Ig-α (fig. 2) [57, 67]. A significant portion of these kinases associated with the resting receptor complex are phosphorylated at their carboxy-terminal tyrosine and therefore inactive. Upon clustering of the receptor, these constitutively associated kinases become activated and capable of phosphorylating the ARH 1 tyrosines of Ig-α and Ig-β. Phosphorylation would change the binding properties of Ig-α and Ig-β such that other Src-family kinases could bind, not via their amino-terminal 10 amino acids [57], but via their SH2 domains. This would enhance the kinase's specific activity by unfolding the kinase to allow its catalytic domain access to substrate [59, 68]. Syk could be recruited in a similar manner although it is unclear how this would enhance its activity.

Once tyrosine phosphorylation of Ig-α and Ig-β occurs, the ARH 1 motifs may function, via other mechanisms to recruit and activate secondary effectors which contain SH2 domains. For example, BCR stimulation most likely activates ras by recruiting it to a complex containing mSOS, Grb2 and Shc [10, 12]. In T cells, this ras activation complex is probably bound to TCRζ via Shc [69] and therefore it is likely that a similar complex binds the BCR.

Although some degree of specialization could occur with phosphotyrosine-specific binding, the crucial residues that determine SH2 binding, the Y + 3 sites, are conserved in the ARH 1 motifs [58]. This suggests that the ARH 1 motifs are capable of binding only to a limited subset of SH2-containing proteins. In particular, as the residues carboxy-terminal to the conserved ARH 1 tyrosines are identical in both Ig-α and Ig-β chains, we predict that upon phosphorylation, there will be identical binding activities for SH2-containing proteins, such as we have demonstrated for Fyn [56].

A Larger Activated Receptor Complex

The activation of tyrosine kinases and other effectors by the BCR does not occur in isolation but is attenuated or enhanced by a variety of coreceptor or cosignal transducer structures. The best studied of these is the tyrosine phosphatase CD45 which is necessary for receptor-initiated signal propagation [70]. The method by which CD45 functions in signal propagation is unclear. One possibility is that CD45 may dephosphorylate the regulatory carboxy-terminal phosphotyrosine of receptor-associated kinases. This process might be facilitated by phosphorylated Ig-α which would displace the negative regulatory phosphotyrosine of the Src-family kinase from its SH2 domain thereby making it accessible to dephosphorylation by CD45 (fig. 2) [71–74]. Dephosphorylation is predicted to stabilize the activated form of the kinase [75]. In this model, which is well supported by the literature and our results, Ig-α acts as a catalyst which stabilizes the transition state of Fyn and other Src-family members, from an inactive to an active kinase.

CD45 may also augment signal transduction by regulating the phosphorylative state of Ig-α and Ig-β. CD45 comodulates with the receptor complex and is capable of dephosphorylating Ig-α/β [70, 76]. In the particular case of Src-family kinases binding to phospho Ig-α or Igβ, we predict that one phosphotyrosine would be accessible to CD45. Dephosphorylation at that residue may facilitate the release of an activated kinase from the receptor complex allowing the kinase better access to substrate. Additionally, the release of an activated kinase from the receptor may allow other kinases to bind the receptor and be activated. Thus, a dynamic balance between kinase and phosphatase activity at the receptor complex may serve to amplify tyrosine kinase activation.

An alternate possibility is that the activated kinases stay associated with the receptor complex and act as bridges to secondary effectors (fig. 3). For instance, while the SH2 domain of Fyn was bound to Ig-α, its SH3 domain may interact with PI3-k and its aminoterminus with PLCγ [16, 77]. In this model, a large activation complex might form in a manner analogous to that which forms upon activation of the platelet-derived growth factor receptor [19, 78]. Effectors which may be bound to such a complex could also form intracytoplasmic bridges to other receptor complexes which may influence signal transduction. For example, PI3-k may not only interact with receptor-bound Lyn or Fyn via its SH3 domain but with CD19 via its two SH2 domains [79] (see below). In the latter case binding to the two transmembrane receptors may be necessary for maximal activation. Although possible in theory, it has been difficult to establish, by immunoprecipitation, that a large activation complex is actually formed upon antigen receptor stimulation [69, 80]. Regardless of how large an activation complex is formed, it is likely that the kinases activated by the BCR interact with and modify coreceptors such as CD19 and CD22 to augment signal transduction.

CD22 may be necessary for BCR-mediated signal transduction in that, in at least one report, naturally occurring B cells lacking CD22 are incompetent to signal via the BCR [81]. The cytoplasmic domain of CD22 contains several tyrosines including two spaced YXXL motifs, similar to the ARH 1 motif [82]. Upon BCR pertubation, the cytoplasmic tail of CD22, probably at the above consensus sites, is almost immediately phosphorylated [83]. These phosphorylated sites may then act to recruit and activate signal transducers as part of a BCR-initiated cascade of activation [84].

In a similar way CD19, which is a component of the CD21/CD19 TAPA 1 (comple-

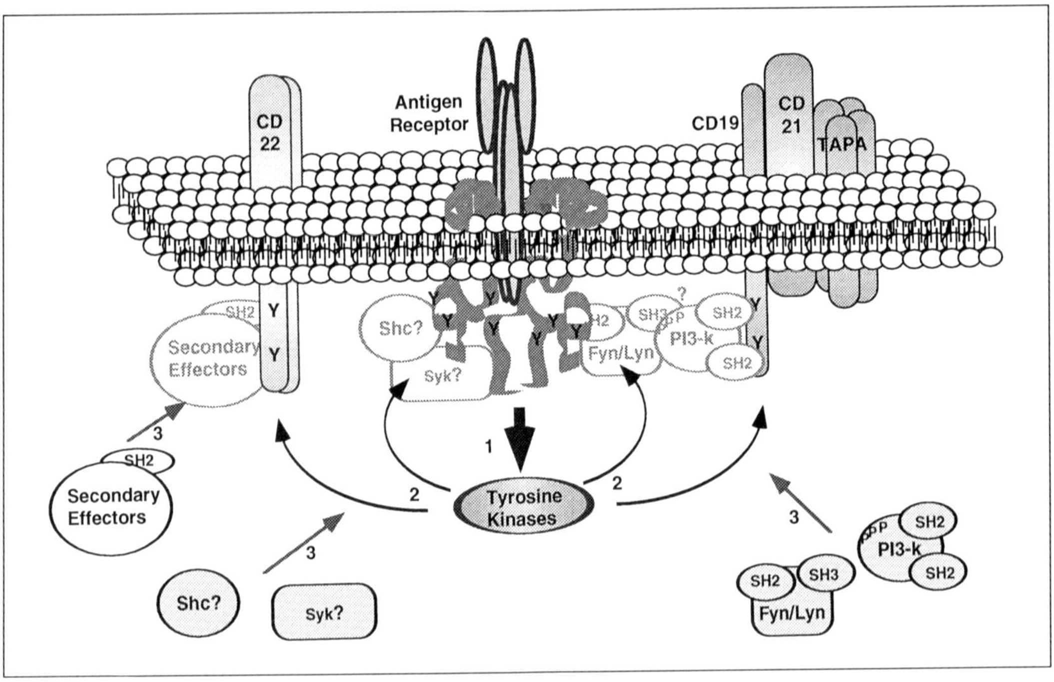

Fig. 3. Establishing a larger activated receptor complex. 1. Perturbation of the receptor leads to the activation of receptor associated tyrosine kinases (see figure 2 and corresponding discussion). 2. These kinases, in turn, phosphorylate Ig-α, Ig-β and directly or indirectly, CD19 and CD22. 3. These phosphotyrosines, and the sequences surrounding them, form recruitment sites for the translocation and activation of a myriad of secondary and tertiary signal transducers.

ment receptor 2) complex, may function in augmenting BCR-mediated signal transduction by providing SH2 domain binding sites [85]. When the complement receptor 2 complex is coaggregated with the BCR, presumably by simultaneous binding of complement-fixed immune complexes, the threshold for BCR stimulation is dramatically decreased [86]. Its cytoplasmic tail contains two YXXM motifs which become phosphorylated following BCR pertubation and act to recruit and activate PI3-k [79]. Since PI3-k may also bind to and be activated by some of the Src-family kinases, PI3-k may serve as an intracellular bridge between the BCR and complement receptor 2 complexes.

Therefore, one can envision a large functional receptor complex in which kinase activation, initiated by the BCR, radiates out from the aggregated antigen receptor complexes to phosphorylate receptor-associated chains which in turn serve as binding and activation sites for a variety of characterized and uncharacterized signal transducers (fig. 3). The separation of kinase activation and secondary signal transducer binding/activating domains is analogous to the insulin receptor with its intrinsic tyrosine kinase activity and its immediate substrate, insulin receptor substrate 1, which upon phosphorylation acts to recruit multiple signal transducers [87].

References

1 Cambier JC, Pleiman CM, Clark MR: Signal transduction by the B cell antigen receptor and its coreceptors. Annu Rev Immunol 1994;12: 457–486.

2 Gold MR, Law DA, DeFranco AL: Stimulation of protein tyrosine phosphorylation by the B-lymphocyte antigen receptor. Nature 1990; 345:810–813.

3 Burkhardt AL, Brunswick M, Bolen JB, Mond JJ: Anti-immunoglobulin stimulation of B lymphocytes activates Src-related protein-tyrosine kinases. Proc Natl Acad Sci USA 1991;88:7410–7414.

4 Campbell MA, Sefton BM: Protein tyrosine phosphorylation is induced in murine B lymphocytes in response to stimulation with anti-immunoglobulin. EMBO J 1990; 9:2125–2131.

5 Carter RH, Park DJ, Rhee SG, Fearon DT: Tyrosine phosphorylation of phospholipase C induced by membrane immunoglobulin in B lymphocytes. Proc Natl Acad Sci USA 1991;88:2745–2749.

6 Coggeshall KM, McHugh JC, Altman A: Predominant expression and activation-induced tyrosine phosphorylation of phospholipase C-γ2 in B lymphocytes. Proc Natl Acad Sci USA 1992;89:5660–5664.

7 Buday L, Downward J: Epidermal growth factor regulates p21(ras) through the formation of a complex of receptor. Grb2 adapter protein, and sos nucleotide exchange factor. Cell 1993;73:611–620.

8 Egan SE, Giddings BW, Brooks MW, Buday L, Sizeland AM, Weinberg RA: Association of sos ras exchange protein with Grb2 is implicated in tyrosine kinase signal transduction and transformation. Nature 1993;363:45–51.

9 McCormick F: Signal transduction – how receptors turn ras on. Nature 1993;363:15–16.

10 Tordai A, Franklin RA, Patel H, Gardner AM, Johnson GL, Gelfand EW: Cross-linking of surface IgM stimulates the Ras/Raf-1/MEK/Mapk cascade in human B lymphocytes. J Biol Chem 1994;269:7538–7543.

11 Stokoe D, Macdonald G, Cadwallader K, Symons M, Hancock JF: Activation of raf as a result of recruitment to the plasma membrane. Science 1994;246:1463–1467.

12 Saxton TM, van Oostveen I, Bowtell D, Aebersold R, Gold MR: B cell antigen receptor cross-linking induces phosphorylation of the p21ras oncoprotein activators SHC and mSOS-1, and a 145-kDa tyrosine-phosphorylated protein. J Immunol 1994;153:623–636.

13 Gulbins E, Coggeshall KM, Baier G, Katzav S, Burn P, Altman A: Tyrosinekinase-stimulated guanine nucleotide exchange activity of vav in T-cell activation. Science 1993;260: 822–825.

14 Bustelo XR, Barbacid M: Tyrosine phosphorylation of the vav protooncogene product in activated B cells. Science 1992;256:1196–1199.

15 Downward J, Graves J, Cantrell D: The regulation and function of p21ras in T cells. Immunol Today 1992;13:89–92.

16 Pleiman CM, Hertz WM, Cambier JC: Activation of phosphatidylinositol-3′ kinase by Src-family kinase SH3 binding to the p85 subunit. Science 1994;263:1609–1612.

17 Rodriguez-Viciana R, Warne PH, Dhand R, Vanhaesebroeck B, Gout I, Fry MJ, Waterfield MD, Downward J: Phosphatidylinositol-3-OH kinase as a direct target of ras. Nature 1994;370:527–532.

18 Liscovitch M, Cantley LC: Lipid second messengers. Cell 1994;77: 329–334.

19 Valius M, Kazlauskas A: Phospholipase-C-gamma-1 and phosphatidylinositol-3 kinase are the downstream mediators of the PDGF receptor's mitogenic signal. Cell 1993; 73:321–334.

20 Klaus GGB: Role of G-proteins in receptor signaling in T and B cells. Semin Immunol 1990;2:151–157.

21 Leprince C, Draves KE, Ledbetter JA, Torres RM, Clark EA: Characterization of molecular components associated with surface immunoglobulin-M in human lymphocyte-B-presence of tyrosine and serine/threonine protein kinases. Eur J Immunol 1992;22:2093–2099.

22 Rogers J, Early P, Carter C, Calame K, Bond M, Hood L, Wall R: Two mRNAs with different 3′ ends encode membrane-bound and secreted forms of immunoglobulin μ chain. Cell 1980;20:303–312.

23 Cheng HL, Blattner FR, Fitzmaurice L, Mushinski JF, Tucker PW: Structure of genes for membrane and secreted IgD heavy chains. Nature 1982;296:410–415.

24 Gold MR, Matsuchi L, Kelly RB, DeFranco AL: Tyrosine phosphorylation of components of the B-cell antigen receptors following receptor crosslinking. Proc Natl Acad Sci USA 1991;88:3436–3440.

25 Campbell KS, Cambier JC: B lymphocyte antigen receptors (mIg) are non-covalently associated with a disulfide-linked, inducibly phosphorylated glycoprotein complex. EMBO J 1990;9:441–448.

26 Hombach J, Leclercq L, Radbruch A, Rajewsky K, Reth M: A novel 34-kd protein coisolated with the IgM molecule in surface IgM-expressing cells. EMBO J 1988;7:3451–3456.

27 Chen J, Stall AM, Herzenberg LA, Herzenberg LA: Differences in glycoprotein complexes associated with IgM and IgD on normal murine B cells potentially enable transduction of different signals. EMBO J 1990;9: 2117–2124.

28 Campbell KS, Hager EJ, Friedrich RJ, Cambier JC: IgM antigen receptor complex contains phosphoprotein products of B29 and mb-1 genes. Proc Natl Acad Sci USA 1991;88:3982–3986.

29 Hombach J, Tsubata T, Leclercq L. Stappert H, Reth M: Molecular components of the B-cell antigen receptor complex of the IgM class. Nature 1990;343:760–762.

30 Clark MR, Friedrich RJ, Campbell KS, Cambier JC: Human Pre-B and B-cell membrane mu-chains are noncovalently associated with a disulfide-linked complex containing a product of the B29-gene. J Immunol 1992;149:2857–2863.

31 Van Noesel CJM, Van Lier RAW, Cordell JL, Tse AGD, Van Schijndel GMW, De Vries EFR, Mason DY, Borst J: The membrane IgM-associated heterodimers on human B cells is newly defined B cell antigen that contains the protein product of the mb-1 gene. J Immunol 1991;146: 3881–3888.
32 Hermanson GG, Eisenberg D, Kincade PW, Wall R: B29: A member of the immunoglobulin gene superfamily exclusively expressed on B-lineage cells. Proc Natl Acad Sci USA 1988;85:6890–6894.
33 Sakaguchi N, Kashiwamura S, Kimoto M, Thalmann P, Melchers F: B lymphocyte lineage-restricted expression of mb-1, a gene with CD3-like structural properties. EMBO J 1988;7:3457–3464.
34 Venkitaraman AR, Williams GT, Dariavach P, Neuberger MS: The B-cell antigen receptor of the five immunoglobulin classes. Nature 1991; 352:777–781.
35 Williams GT, Venkitaraman A, Gilmore D, Neuberger M: The sequence of the μ transmembrane segment determines the tissue specificity of the transport of immunoglobulin M to the cell surface. J Exp Med 1990;171:947–952.
36 Parikh VS, Bishop GA, Liu KJ, Do BT, Ghosh MR, Kim BS, Tucker PW: Differential structure-function requirements of the transmembranal domain of the B-cell antigen receptor. J Exp Med 1992;176:1025–1031.
37 Weiss A, Littman DR: Signal transduction by lymphocyte antigen receptors. Cell 1994;76:263–274.
38 Irving BA, Weiss A: The cytoplasmic domain of the T cell receptor zeta chain is sufficient to couple to receptor-associated signal transduction pathways. Cell 1991;64:891–901.
39 Sanchez M, Misulovin Z, Burkhardt AL, Mahajan S, Costa T, Franke R, Bolen JB, Nussenzweig M: Signal transduction by immunoglobulin is mediated through Ig-alpha and Ig-beta. J Exp Med 1993;178:1049–1055.
40 Burkhardt AL, Costa T, Misulovin Z, Stealy B, Bolen JB, Nussenzweig MC: Ig-alpha and Ig-beta are functionally homologous to the signaling proteins of the T-cell receptor. Mol Cell Biol 1994;14:1095–1103.
41 Choquet D, Ku G, Cassard S, Malissen B, Korn H, Fridman WH, Bonnerot C: Different patterns of calcium signaling triggered through two components of the B lymphocyte antigen receptor. J Biol Chem 1994;269:6491–6497.
42 Law DA, Chan VWF, Datta SK, DeFranco AL: B-cell antigen receptor motifs have redundant signaling capabilities and bind the tyrosine kinases PTK72, Lyn and Fyn. Curr Biol 1993;3:645–657.
43 Kim KM, Alber G, Weiser P, Reth M: Differential signaling through the Ig-alpha and Ig-beta components of the B-cell antigen receptor. Eur J Immunol 1993;23:911–916.
44 Reth M: Antigen receptor tail clue. Nature 1989;338:383–384.
45 Keegan AD, Paul WE: Multichain immune recognition receptors: Similarities in structure and signaling pathways. Immunol Today 1992;13: 63–68.
46 Cambier JC, Bedzyk W, Campbell K, Chien N, Friedrich J, Harwood A, Jensen W, Pleiman C, Clark MR: The B-cell antigen receptor: Structure and function of primary, secondary, tertiary and quaternary components. Immunol Rev 1993; 132:85–106.
47 Clark MR, Campbell KS, Kazlauskas A, Johnson SA, Hertz M, Potter TA, Pleiman C, Cambier JC: The B-cell antigen receptor complex: Association of Ig-alpha and Ig-beta with distinct cytoplasmic effectors. Science 1992;258:123–126.
48 Letourneur F, Klausner RD: Activation of T cells by a tyrosine kinase activation domain in the cytoplasmic tail of CD3 epsilon. Science 1992;255:79–82.
49 Letourneur F, Klausner RD: T-cell and basophil activation through the cytoplasmic tail of T-cell-receptor zeta family proteins. Proc Natl Acad Sci USA 1991;88:8905–8909.
50 Romeo C, Amiot M, Seed B: Sequence requirements for the induction of cytolysis by the T cell antigen/Fc receptor zeta chain. Cell 1992;68:889–897.
51 Wegener A-MK, Letourneur F, Hoeveler A, Brockeker T, Luton F, Malissen B: The T cell receptor/CD3 complex is composed of at least two autonomous transduction modules. Cell 1992;68:83–95.
52 Irving BA, Chan AC, Weiss A: Functional characterization of signal transducing motif present in the T cell antigen receptor. J Exp Med 1993;177:1093–1103.
53 Flaswinkel H, Reth M: Dual role of the tyrosine activation motif of the Ig-alpha protein during signal transduction via the B cell antigen receptor. EMBO J 1994;13:83–89.
54 Love PE, Shores EW, Lee EJ, Grinberg A, Munitz TI, Westphal H, Singer A: Differential effects of zeta and eta transgenes on early alpha/beta T cell development. J Exp Med 1994; 179:1485–1494.
55 Jouvin MHE, Adamczewski M, Numerof R, Letourneur O, Valle A, Kinet JP: Differential control of the tyrosine kinases Lyn and Syk by the two signaling chains of the high affinity immunoglobulin epsilon receptor. J Biol Chem 1994;269: 5918–5925.
56 Clark MR, Johnson SA, Cambier JC: Analysis of Ig-alpha tyrosine kinase interaction reveals two levels of binding specificity and tyrosine phosphorylated Ig-alpha stimulation of Fyn activity. EMBO J 1994; 13:1911–1919.
57 Pleiman CM, Abrams C, Gauen LT, Bedzyk W, Jongstra J, Shaw AS, Cambier JC: Distinct p53/56lyn and p59fyn domains associate with non-phosphorylated and phosphorylated Ig-α. Proc Natl Acad Sci USA 1994; 91:4268–4272.
58 Sonyang Z, Shoelson SE, Chaudhuri M, Gish G, Pawson T, Hasert WG, King F, Roberts T, Ratnofshy S, Lecheider RJ, Neel BG, Birge RB, Fajardo JE, Chou M, Hanafusa H, Schaffhausen B, Cantley LC: SH2 domains recognize specific phosphopeptide sequences. Cell 1993;72: 767–778.
59 Cantley LC, Auger KR, Carpenter C, Duckworth B, Graziani A, Kapeller R, Soltoff S: Oncogenes and signal transduction. Cell 1991;64: 281–302.

60 Eck MJ, Atwell SK, Shoelson SE, Harrison SC: Structure of the regulatory domains of the Src-family tyrosine kinase Lck. Nature 1994;368: 764–769.
61 Yamanashi Y, Kakiuchi T, Mizuguchi J, Yamamoto T, Toyoshima K: Association of B cell antigen receptor with protein tyrosine kinase Lyn. Science 1991;251:191–194.
62 Timson Gauen LKT, Zhu Y, Letoureneur F, Hu Q, Bolen JB, Matis LA, Klausner RD, Shaw A: Interactions of p59fyn and ZAP-70 with T-cell receptor activation motifs: Defining the nature of a signalling motif. Mol Cell Biol 1994;14:3729–3741.
63 Chan AC, Iwashima M, Turck CW, Weiss A: ZAP-70: A 70 kd protein-tyrosine kinase that associates with the TCR zeta-chain. Cell 1992;71: 649–662.
64 Iwashima M, Irving BA, Vanoers NSC, Chan AC, Weiss A: Sequential interactions of the TCR with 2 distinct cytoplasmic tyrosine kinases. Science 1994;263:1136–1139.
65 Couture C, Baier G, Oetken C, Williams S, Telford D, Cardine AM, Bitterlich GB, Fischer S, Burn P, Altman A, Mustelin T: Activation of p56lck by p72syk through physical association and N-terminal tyrosine phosphorylation. Mol Cell Biol 1994;14:5249–5258.
66 Couture C, Baier G, Altman A, Mustelin T: p56lck-independent activation and tyrosine phosphorylation of p72syk by T-cell antigen receptor/CD3 stimulation. Proc Natl Acad Sci 1994;91:5301–5305.
67 Timson-Gauen LK, Kong A-NT, Samelson LE, Shaw AS: The p59Fyn tyrosine kinase associated with multiple T cell receptor subunits through its unique amino-terminal domain. Mol Cell Biol 1992;12: 5438–5446.
68 Carpenter G: Receptor tyrosine kinase substrates – Src homology domains and signal transduction. FASEB J 1992;6:3283–3289.
69 Ravichandran KS, Lee KK, Songyang Z, Cantley LC, Burn P, Burakoff SJ: Interaction of Shc with the zeta chain of the T cell receptor upon T cell activation. Science 1993;262:902–905.
70 Justement LB, Campbell KS, Chien NC, Cambier JC: Regulation of B cell antigen receptor signal transduction and phosphorylation by CD45. Science 1991;252:1839–1842.
71 Hurley TR, Hyman R, Sefton BM: Differential effects of expression of the CD45 tyrosine protein phosphatase on the tyrosine phosphorylation of the Lck, Fyn, and c-Src tyrosine protein kinases. Mol Cell Biol 1993; 13:1651–1656.
72 McFarland EDC, Hurley TR, Pingel JT, Sefton BM, Shaw A, Thomas ML: Correlation between Src family member regulation by the protein-tyrosine-phosphatase CD45 and transmembrane signaling through the T-cell receptor. Proc Natl Acad Sci USA 1993;90:1402–1406.
73 Koretzky GA: Role of the CD45 tyrosine phosphatase in signal transduction in the immune system. FASEB J 1993;7:420–426.
74 Sieh M, Bolen JB, Weiss A: CD45 specifically modulates binding of Lck to a phosphopeptide encompassing the negative regulatory tyrosine of Lck. EMBO J 1993;12:315–321.
75 Shiroo M, Goff L, Biffen M, Shivnan E, Alexander D: CD45-tyrosine phosphatase-activated p59fyn couples the T-cell antigen receptor to pathways of diacylglycerol production, protein kinase-C activation and calcium influx. EMBO J 1992; 11:4887–4897.
76 Lin J, Brown VK, Justement LB: Regulation of basal tyrosine phosphorylation of the B cell antigen receptor complex by the protein tyrosine phosphatase, CD45. J Immunol 1992;149:3182–3190.
77 Pleiman CP, Clark MR, Timson-Gauen LK, Winz S, Coggeshall KM, Johnson GL, Shaw AS, Cambier JC: Mapping of sites on the Src family protein tyrosine kinases p55blk, p59fyn, and p56lyn which interact with the effector molecules phospholipase c-gamma 2, microtubule-associated protein kinase, GTPase-activating protein, and phosphatidylinositol 3-kinase. Mol Cell Biol 1993;13:5877–5887.
78 Williams LT: Signal transduction by the platelet-derived growth factor receptor. Science 1989;243:1564–1570.
79 Tuveson DA, Carter RH, Soltoff SP, Fearon DT: CD19 of B-cells as a surrogate kinase insert region to bind phosphatidylinositol 3-kinase. Science 1993;260:986–989.
80 Exley M, Varticovski L, Peter M, Sancho J, Terhorst C: Association of phosphatidylinositol 3-kinase with a specific sequence of the T cell receptor chain is dependent on T cell activation. J Biol Chem 1994;269: 15140–15146.
81 Pezzutto A, Rabinovitch PS, Dorken B, Moldenhauer G, Clark EA: Role of the CD22 human B cell antigen in B cell triggering by anti-immunoglobulin. J Immunol 1988; 140:1791–1795.
82 Stamenkovic I, Seed B: The B cell antigen CD22 mediates monocyte and erythrocyte adhesion. Nature 1990;345:74–77.
83 Schulte RJ, Campbell MA, Fischer WH, Sefton BM: Tyrosine phosphorylation of CD22 during B-cell activation. Science 1992;258:1001–1004.
84 Clark EA: CD22, a B cell-specific receptor, mediates adhesion and signal transduction. J Immunol 1993; 150:4715–4718.
85 Fearon DT: The CD19-CR2-TAPA-1 complex, CD45 and signaling by the antigen receptor of B lymphocytes. Curr Opin Immunol 1993;5: 341–348.
86 Carter RH, Tuveson DA, Park DJ, Rhee SG, Fearon DT: The CD19 complex of B lymphocytes: Activation of phospholipase C by a protein tyrosine kinase-dependent pathway that can be enhanced by the membrane IgM complex. J Immunol 1991;147:3663–3671.
87 Ando A, Yonezawa K, Gout I, Nakata T, Ueda H, Hara K, Kitamura Y, Noda Y, Takenawa T, Hirokawa N, Waterfield MD, Kasuga M: A complex of Grb2-dynamin binds to tyrosine-phosphorylated insulin receptor substrate-1 after insulin treatment. EMBO J 1994;13:3033–3038.

Author Index Vol. 13, 1994

Subject Index Vol. 13, 1994

Vol. 13, 1994

Immunologic Research

A Selective Reference to Current Research and Practice

Formerly 'Survey of Immunologic Research'

KARGER

S. Karger
Medical and Scientific Publishers
Basel · Freiburg · Paris · London
New York · New Delhi · Bangkok
Singapore · Tokyo · Sydney

Drug Dosage
The authors and the publisher have exerted every effort to ensure that drug selection and dosage set forth in this text are in accord with current recommendations and practice at the time of publication. However, in view of ongoing research, changes in government regulations, and the constant flow of information relating to drug therapy and drug reactions, the reader is urged to check the package insert for each drug for any change in indications and dosage and for added warnings and precautions. This is particularly important when the recommended agent is a new and/or infrequently employed drug.

Printed in Switzerland on acid-free paper by
Thür AG Offsetdruck, Pratteln

Contents Vol. 13, 1994

No. 1

No. 2–3

Molecular Aspects of the Rheumatic Diseases
Proceedings of the 2nd Symposium
Guest Editors: William V. Williams; David B. Weiner, Philadelphia, Pa., USA

No. 4

Immunology: A University of Chicago Perspective

Guest Editor: J.A. Bluestone, Chicago, Ill., USA

Presents exciting new developments

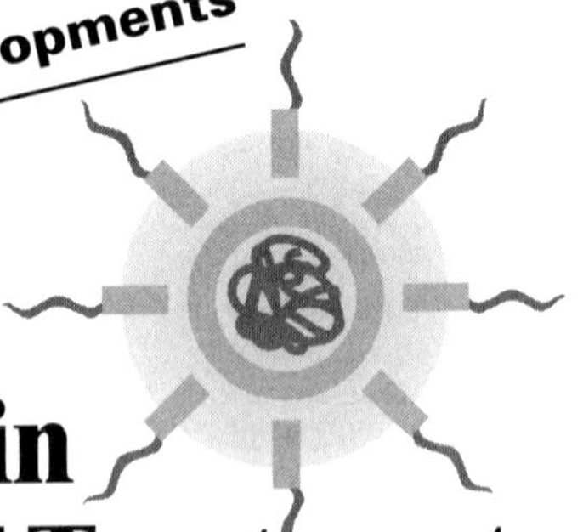

Advanced Technologies in Research, Diagnosis and Treatment of AIDS and in Oncology

Editors: Gaetano Giraldo, Marco Salvatore, Luigi Chieco-Bianchi, Elke Beth-Giraldo

This volume contains a state-of-the-art overview of some of the most exciting new developments in the pathogenesis, immunology and molecular biology of AIDS and in cancer research.
Biomedical researchers, physicians (especially pediatricians and oncologists working in the area), and anyone wishing to keep up to date with the latest developments in the field of AIDS and malignancies will find this book to be a valuable addition to their libraries.

Contents

Antibiotics and Chemotherapy, Vol. 46
Series Editor: Schönfeld, H. (Grenzach)
ISSN 0066–4758

Fields of Interest: Pharmacology, Pharmaceutics; Immunology; Oncology; Infectious Diseases, Vaccination, Virology

Advanced Technologies in Research, Diagnosis and Treatment of AIDS and in Oncology
International Workshop, Naples, October 1993
Editors: Giraldo, G.; Salvatore, M. (Naples); Chieco-Bianchi, L. (Padua); Beth-Giraldo, E. (Naples)
VIII + 168 p., 41 fig., 2 cpl., 26 tab., hard cover, 1994
SFr. 190.– / DM 228.– / US$ 165.25
Prices subject to change
DM price for Germany, US$ price for USA only
ISBN 3–8055–5970–4

KI 94557

KARGER